WORKED EXAMPLES
A LEVEL

BIOLOGY

by
J. Caswell

CELTIC REVISION AIDS

CELTIC REVISION AIDS
Lincoln Way, Windmill Road,
Sunbury on Thames, Middlesex

First published in this edition 1979
Reprinted 1980

ISBN 017 751153 2

Printed in Hong Kong

CONTENTS

1. Describe the process of conjugation in Paramecium and discuss its significance.

Reproduction in Paramecium occurs in two ways. One way is called transverse fission and the second, conjugation. Conjugation is a link between the higher Protozoa and those of the Metazoa. Two Paramecia come together by a temporary adhesion by their ventral surfaces. These Paramecia are called the conjugants. In Paramecia caudatum the conjugants must be from different strains. The micronucleus becomes very active and separates from the meganucleus. The meganucleus gradually disintegrates and disappears into the endoplasm. The micronucleus of each individual increases in size and divides twice forming four nuclei. Three of these nuclei abort and during the divisions of the nuclei there is a reduction in their chromosomes to the haploid number. The remaining nucleus then divides again to produce two gametic nuclei of different potentialities. These act as gametes and from each paramecium one nucleus passes over into the other where it fuses with the stationary nucleus to form a zygotic nucleus. The

Nuclear changes during conjugation

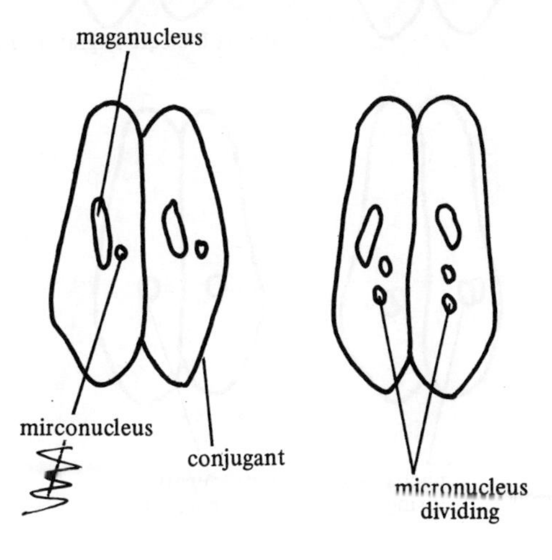

conjugants then separate. The zygotic nucleus undergoes repeated divisions forming eight nuclei and then the body divides transversely. Each conjugant contains four daughter nuclei. Then there is further division so that there are produced four new individuals each of which has two nuclei. One enlarges to become the meganucleus and the other forms the micronucleus. As conjugation involves a fusion of nuclear material it is considered to be a form of sexual reproduction.

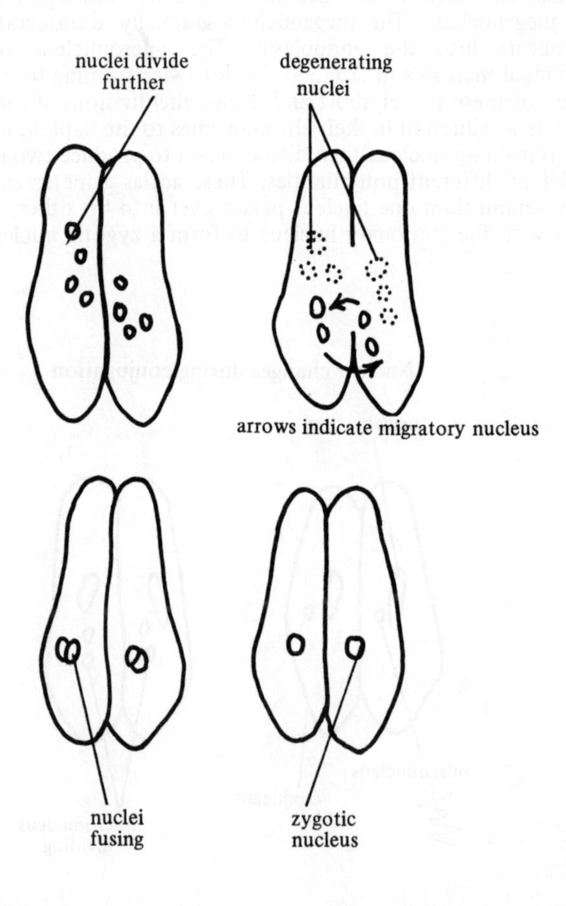

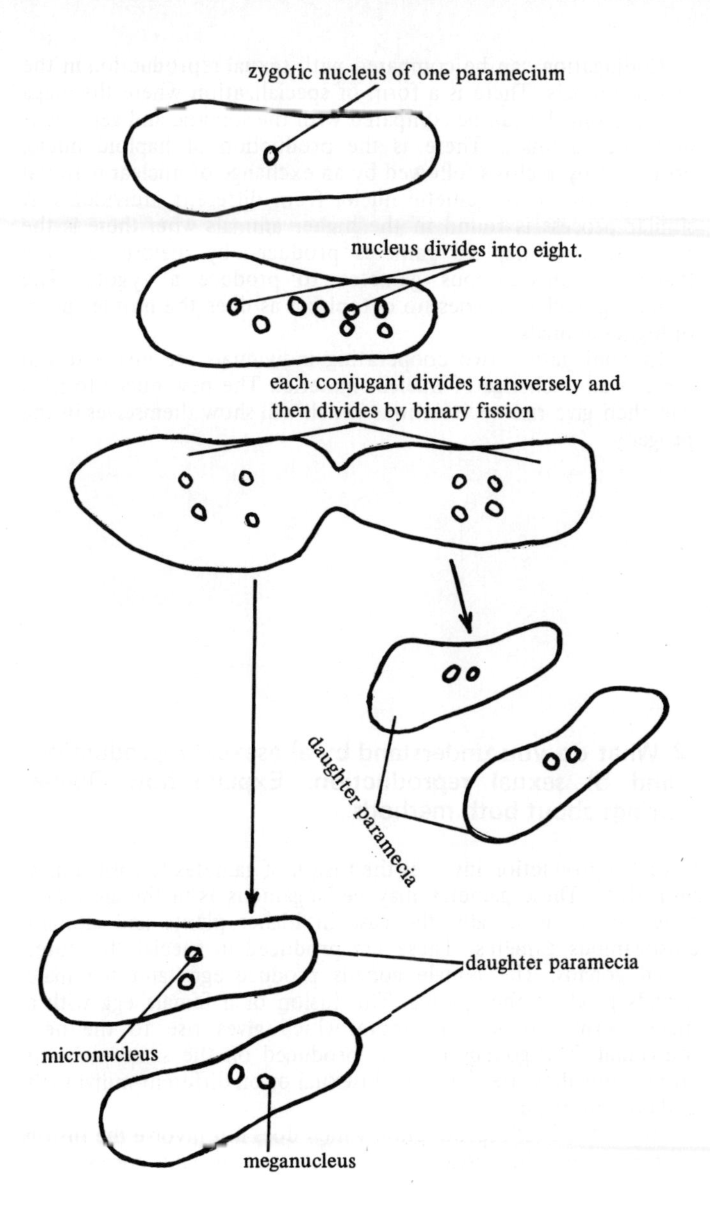
zygotic nucleus of one paramecium
nucleus divides into eight.
each conjugant divides transversely and
then divides by binary fission
daughter paramecia
daughter paramecia
micronucleus
meganucleus

Conjugation can be compared with sexual reproduction in the higher animals. There is a form of specialization where the mega and micronuclei can be compared with the somatic and germ cells in higher animals. There is the production of haploid nuclei produced by meiosis followed by an exchange of nuclear material with the fusion of gametic nuclei from different individuals. A similar process is found in the higher animals wher there is the production of haploid gametes produced by meiosis and the fusion of anisogamous gametes to produce a zygote. The migrating nucleus carries no cytoplasm as does the motile sperm of higher animals.

In conjugation two cooperating individuals are involved and there is an exchange of nuclear material. The new nuclei formed will then give rise to variations which will show themselves in the progeny.

2. What do you understand by a) asexual reproduction and b) sexual reproduction. Explain how Obelia brings about both methods.

Sexual reproduction involves the fusion of gametes to form a new individual. These gametes may be isogamous as in the algae and fungi, or as is usually the case in higher plants and animals anisogamous gametes. These are produced in special structures called gonads. The female gonads produce eggs and the male gonads produce the sperms. The fusion of a female egg with a male sperm produces a zygote which gives rise to the new individual. The gonads may be produced on the same plant or animal and they are therefore bisexual or on different individuals and are unisexual.

All methods of reproduction which does not involve the fusion of gametes is called asexual reproduction. It usually corresponds very closely to vegetative propagation, where a separated part of the plant or animal is used to produce a new one. e.g. Hydra and the gemma of mosses. Asexual reproduction is also exhibited by the spore producing processes exhibited by many individuals e.g. chlamydomonas and the mosses and liverworts.

Obelia is a marine coelenterate attached to the surface of seaweeds. They occur in large fixed colonies which are very branched and so sway in the water with the small polyps protruding from the cups at the ends of the branches. Asexual reproductive zooids called blastostyles are found in the axils of the lower branches. The blastostyles have neither tentacles nor a mouth for they are purely reproductive zooids. Each consists of a club shaped hollow column. Medusae are budded off in succession from the apex downwards and therefore all the stages in their development may be seen in a single blastostyle. The medusae are small and free living and are released during the Spring and Summer.

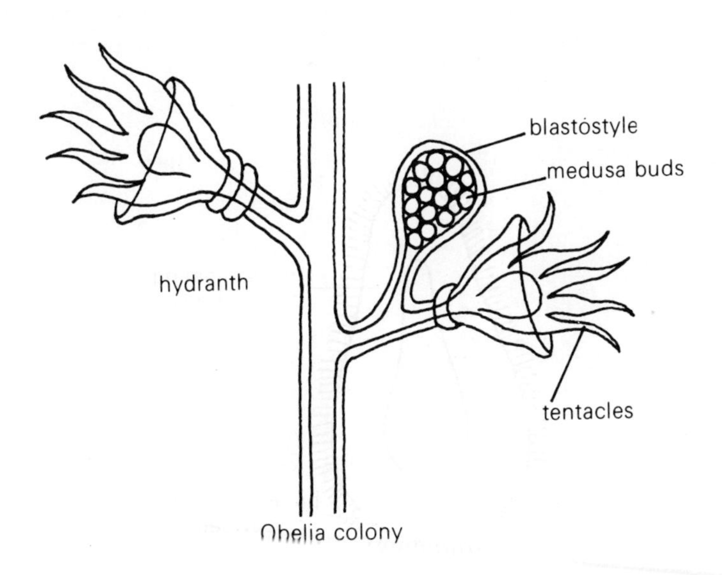

Obelia colony

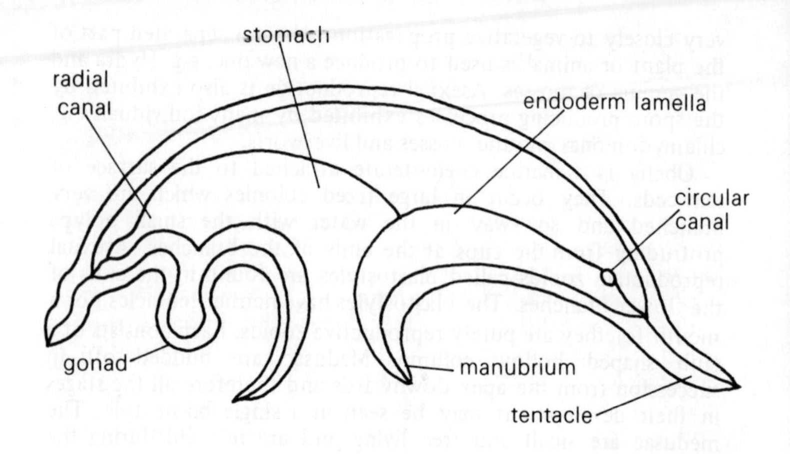

L. S. of Obelia medusa

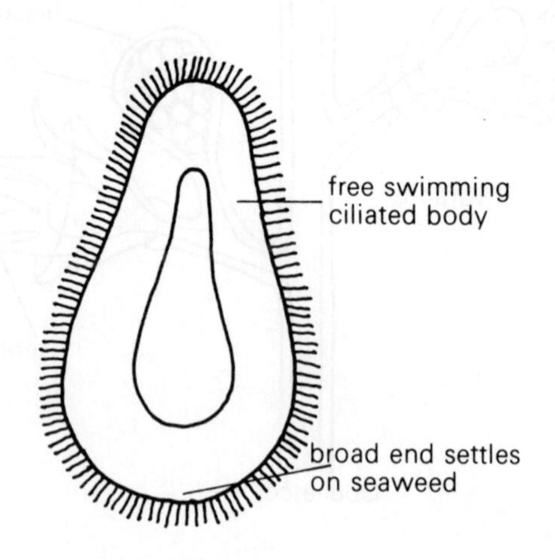

Planula larva

The young medusa is expelled from the pore by currents produced by the rhythmical contractions of the hydranth. The medusa is like a transparent shallow saucer, the middle of the concave surface being produced into a short manubrium. The rim of the medusa bell is furnished with tentacles. It has four radial canals running from the gastric cavity to the circular canal. On the course of the radial canals and at the end of a short branch are globose sacs which are the gonads. The medusae are unisexual, thus the gonads produce sperms in male and ova in females.

When the gonads are mature the ripe ova are shed into the water by the rupture of the sacs and fertilization takes place. The resulting zygote develops into a hollow blastula and eventually an elongated planula larva is formed. It is ciliated and swims freely for a time, then settling by its broader end while the other end develops a mouth and tentacles surrounding it. From this the rest of the colony is developed. The medusae and the planula larva are essentially formed for dispersal.

Both fixed and free forms are diploid, the haploid phase is represented by the gametes.

The colonial form produced from the zygote via the planula larva produces many hydranths before the blastostyles are developed and hence the medusae. The hydranths may be regarded as asexual buds which do not become detached. The colonial form does not produce any gametes. In Obelia there are asexual and sexual zooids.

3. What features would enable you to recognise Fasciola as a) belonging to the Phylum Platyhelminthes, b) a parasite.

The features that recognise Fasciola as belonging to the Phylum Platyhelminthes are as follows:-

1. Bilateral symmetry.

The various organs of the body are symmetrically disposed with regard to the mid-line so that a plane passing through this mid-line at right angles to the dorsal and ventral surfaces bisects the animal into similar right and left halves. Fasciola is bilaterally symmetrical.

2. A triploblastic metazoan

The development of a thin layer of cells, the mesoderm, lying between the ectoderm and endoderm provides an increase in size and complexity. Fasciola is triploblastic.

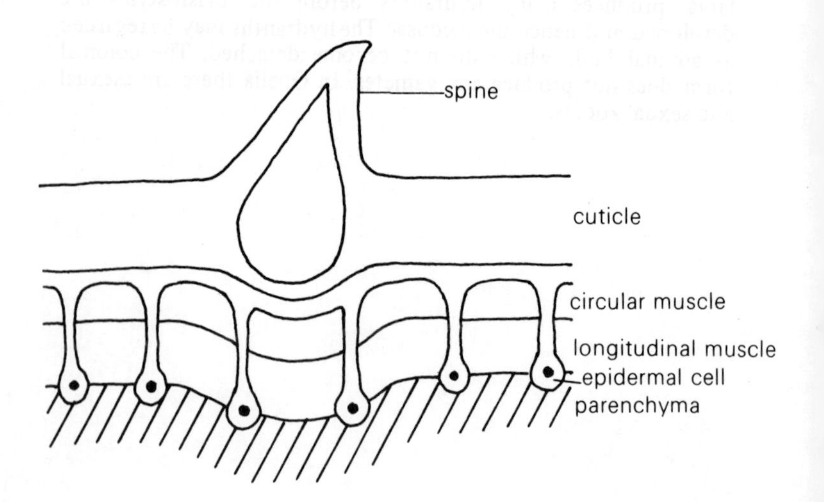

3. Single opening of the alimentary canal.

The mouth lies in the middle of the anterior sucker. It opens into a short suctorial pharynx. The oesophagus is also short and connects with the intestine which has two main branches and numerous branched caeca running throughout the body.

4. Excretory system.

The posterior median excretory canal opens at the terminal pore. Throughout its length it receives many large ducts which in turn drain smaller tubules whose ultimate branches end in flame cells.

5. The reproductive organs are hermaphrodite.

Fasciola is hermaphrodite and the reproductive organs are very complicated. The testes lie one behind the other in the posterior half of the body. From the centre of the testis, a fine vas deferens leads forwards and two vasa deferentia unite close to the ventral sucker to form a wider vesicula seminalis. This passes forwards as an ejaculating duct through the protrusible penis housed in a muscular sac. Where the duct enters the penis sac, it is surrounded by prostate cells.

The ovary is much branched and thicker. It lies anteriorly on the right hand side of the body and from it an oviduct leads into the mid-line. Numerous vitelline glands lie laterally and their products are collected by a series of fine canals which lead to the two main ducts. These pass inwardly and meet in a small vitelline reservoir from which a short duct passes forward to join the oviduct. At their junction a narrow Laurer's canal leads dorsally and opens on the body surface. From the junction of the oviduct, vitelline duct and Laurer's canal, the eggs pass forward into a small reservoir surrounded by glandular cells which secrete a substance which hardens the shells. The eggs then pass into a wide convoluted uterus which opens near the penis into the genital atrium.

6. Flatworms.

The body of Fasciola is dorsiventrally flattened and very thin. It has a more or less oval outline.

All these characters are characteristics of the Phylum Platyhelminthes.

The parasitic features of Fasciola are its thin leaf-like body with a thick cuticle which protects the animal from anti-toxins in the host's body. Fasciola also bears suckers, an anterior and ventral sucker which attach the animal firmly to the host. There are no cilia or sense organs and the animal produces large numbers of eggs.

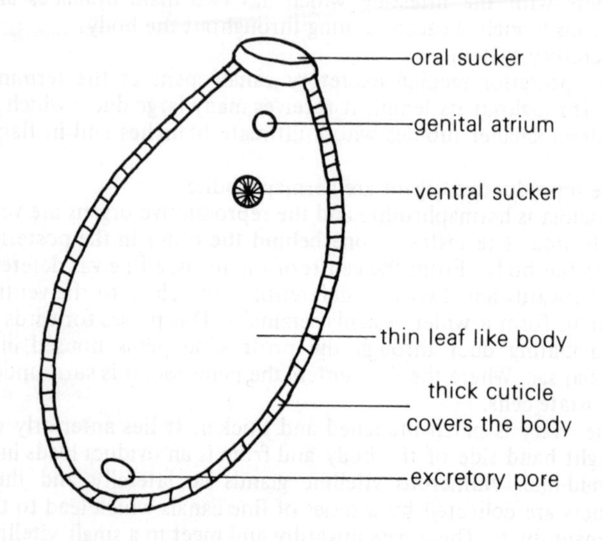

Fasciola ventral view

4. Show how the cells in the body wall of an earthworm receive from the external environment a) a supply of food and b) a supply of oxygen.

Earthworms burrow through the soil and it is from here that they obtain their food. The food is mainly vegetable matter which is sucked into the thick walled pharynx by a pumping action. The alimentary canal runs throughout the length of the body beginning at the mouth. This is a small crescent shaped structure bordered by the peristomium and overhung by the prostomium. The mouth leads into a small chamber, the buccal cavity and then the pharynx. Before the food can be ingested it must be moistened and softened. This occurs when mucin containing proteolytic enzymes is secreted onto the food. Once the food has been sucked into the pharynx it is forced backwards along the thin walled oesophagus, where it come into contact with the milky secretions of the chalk producing glands. These milky secretions

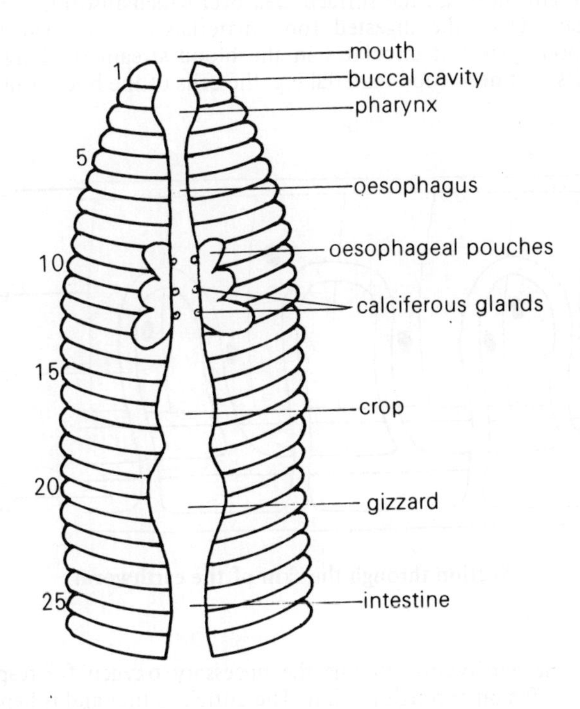

Digestive system of earthworm

neutralize any acids that are present in the food. The partly digested food material is stored in the crop. The next part of the canal is the thick hard muscular gizzard. In the gizzard the food is crushed with small sharp stones and then churned around. The main digestive process takes place in the intestine. Digestive juices produced by the gland cells contain proteolytic, amylolytic and lipolytic enzymes. The intestine is also used for the absorption of food materials into the blood capillaries. The dissolved substances pass easily through the walls and the typhosole which hangs from the gut increases the surface area over which absorption can take place. Once the digested food materials are absorbed into the blood system it circulates in the blood stream reaching all the cells that need food material e.g. the cells of the body wall.

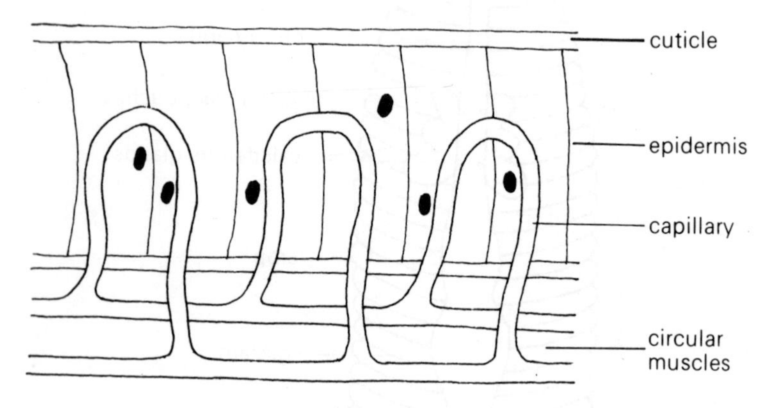

Section through the skin of the earthworm

The earthworm obtains the necessary oxygen for respiration by diffusion through its skin. The cuticle is thin and is kept moist by secretions from the epidermal mucous glands and from fluid from the dorsal pores. The epidermis is richly supplied by a looped blood system which comes very near the surface. The blood contains the respiratory pigment haemoglobin which enters into chemical combination with the oxygen to form oxy-haemoglobin. The ratio of surface area to volume remains constant because of the elongated shape of the body. The oxyhaemoglobin circulates with the blood and releases the oxygen to the cells where there is a low oxygen tension. It must be emphasised that the blood does not come in direct contact with the tissue cells and that oxygen is passed on through a tissue fluid intermediary. Thus the oxygen reaches the cells of the body through the skin and hence by circulation in the blood.

5. By means of the external features of the cockroach, describe the chief characteristics of insects.

The cockroach is classified as an insect and insects have the following external characteristics. The body is divided into three sections, the head, thorax and abdomen. The head represents six segments though all external signs of segmentation have been lost. The thorax has three segments and the abdomen eleven segments.

The head bears a pair of long antennae which are concerned with the sense of smell. A pair of large compound eyes are found on either side of the head and surrounding the mouth are a set of mouth parts, the mandibles, maxillae and the labium.

The three segments of the thorax are the prothorax, the mesothorax and the metathorax respectively. On each segment a pair of walking legs occur. On the mesothorax and metathorax a pair of wings are found. The forewings are heavily thickened to act as protective covers for the membranous hind wings.

The abdomen shows the segmentation of the body although not all eleven segments are recognisable externally. Along both sides of the body can be found a series of openings called spiracles which are the openings of the breathing pores. The terminal region of the abdomen is modified in connection with the reproductive system. In the male two slender processes, the anal styles are present, while in the female the sternites of the seventh segment are prolonged backwards to form a keel like structure in which lies the opening to the reproductive system. A pair of tail feelers called the cerci are the last pair of appendages and they serve as posterior antenna which are sensitive to touch and vibrations.

All these external features describe the chief characteristics of the insects.

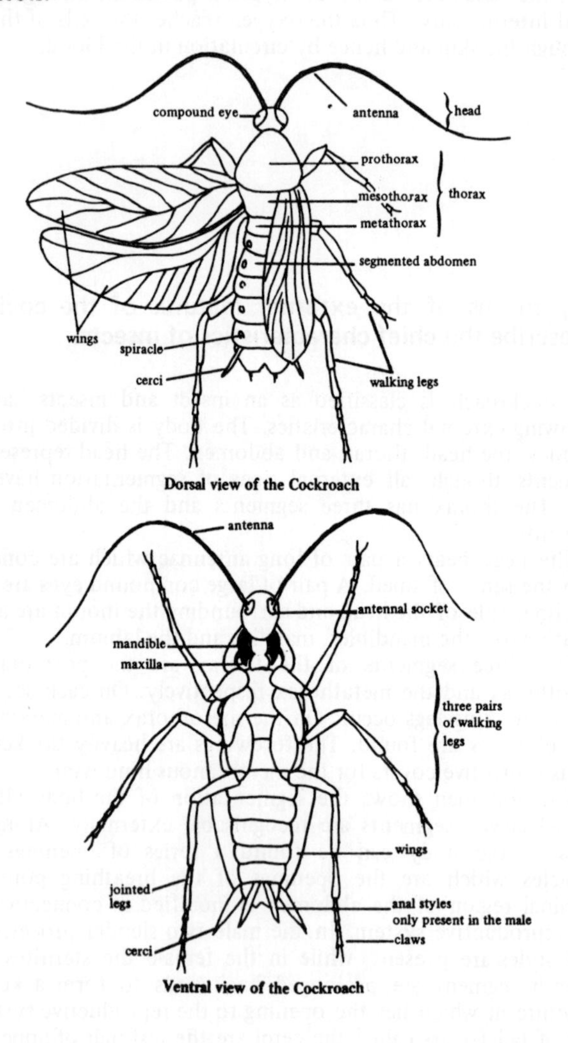

Dorsal view of the Cockroach

Ventral view of the Cockroach

6.Explain the structure and function of the following: nomatocyst, gametocyst, sporocyst, statocyst and trichocyst.

Nematocyst

The body wall of Hydra is composed of a variety of cells. The interstial cells lie between the musculo-epithelial cells. These have the power of differentiating into ectodermal cells and they frequently replace the nematoblasts.

The nematoblasts are specialised cells lying in groups embedded in the superficial layers of the ectoderm. Each nematoblast secretes a **nematocyst**. This is used for defence or for immobilising animals or killing the animals on which Hydra feeds. Some nematocysts are also involved with movement.

There are four kinds of nematocysts, the penetrants, the volvents, the large glutinants and small glutinants.

Each penetrant is composed of a capsule with a lid. There is a short wide tube inverted into the capsule and differentiated into a shaft and a spinneret with barbs. It also bears a coiled filament which lies in a fluid filled capsule and winds around the shaft. On one side of the lid there projects a small sensory bristle the cnidocil. On the capsule wall are retractile rods which are attached to the contractile threads running to the base of the nematocyst.

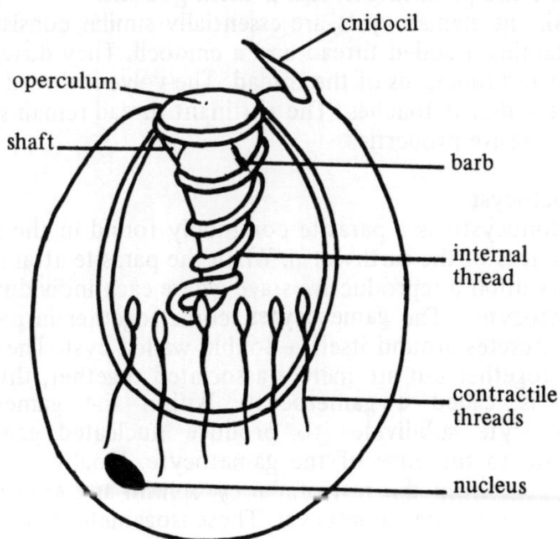

A nematocyst of Hydra before discharge

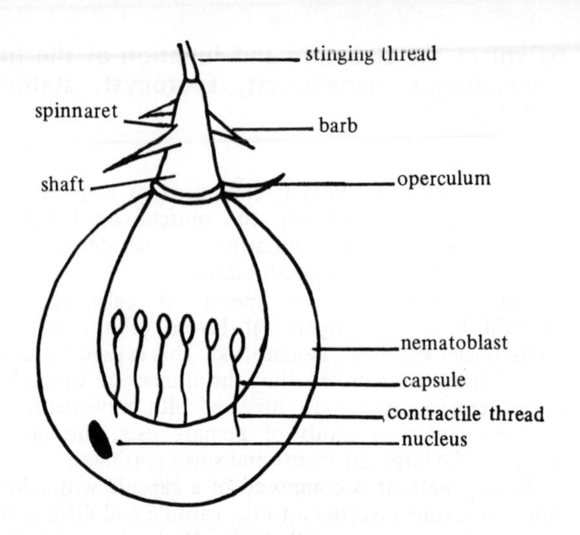

A nematocyst of Hydra after discharge

When the cnidocil is triggered by a small animal, a long thread turns inside out and shoots out from the capsule penetrating the animal and paralising it with protein poisons.

All the nematocysts are essentially similar consisting of a sac containing a coiled thread and a cnidocil. They differ only in the form and functions of the thread. The volvent thread coils around objects that it touches. The glutinant thread remains straight and has adhesive properties.

Gametocyst

Monocystis is a parasite commonly found in the reproductive apparatus of the earthworm. When the parasite attains maturity it enters upon a reproductive stage where each individual becomes a gametocyte. **The gametocytes** come together in pairs and each pair secretes around itself a double walled cyst. The cysts do not fuse together but are merely associated together, this association cyst is called a **gametocyst.** Within the gametocyst each gametocyte subdivides to produce nucleated gametes. These migrate to the edge of the gametocyte. Finally the gametes are separated from the rest of the cytoplasm and are set free inside the cavity of the gametocyst. These isogametes fuse i.e. one from each gametocyte to form zygotes. These develop into sporoblasts

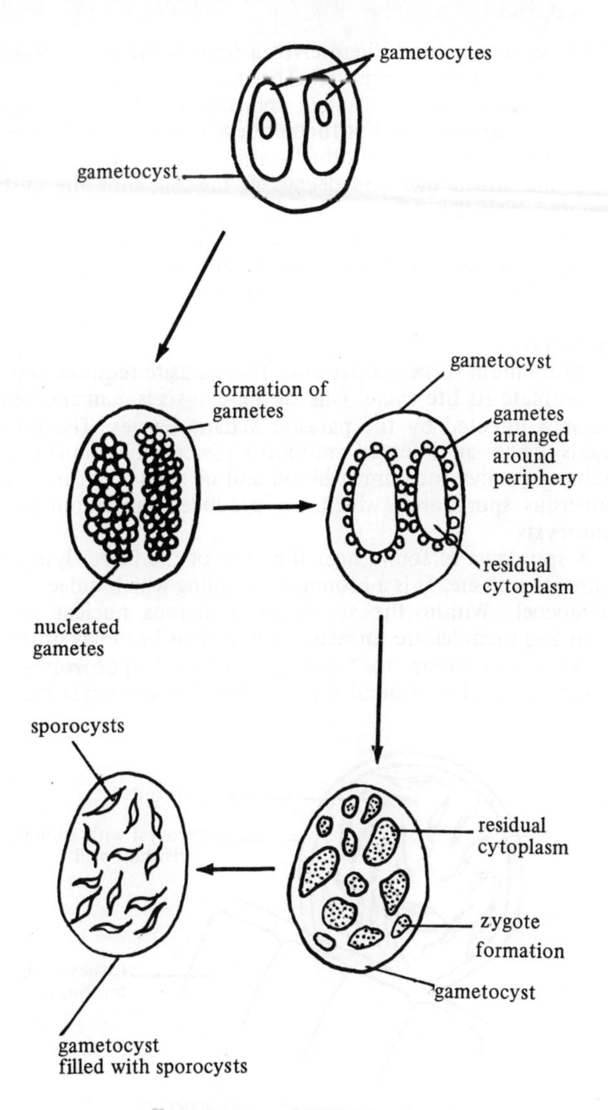

Gametocyst of Monocystis

which secrete around themselves a hard resistant covering, the sporocyst or pseudonavicella. The sporocyst divides producing eight sickle-shaped nucleated sporozoites. All are contained within the gametocyst. For further development to take place the sporozoites must reach another worm. This can only be achieved when the worm dies and decays in the soil, thus the cysts are liberated into the soil or when the worm is eaten by a bird and the cysts are liberated in the faeces. The cyst walls are then digested by the digestive juices and eventually infest the reproductive apparatus of the earthworm.

Sporocyst

Plasmodium vivax is a parasite. This parasite requires two hosts to complete its life cycle. One of these hosts is man and when he becomes infected by the parasite Malaria ensues. The infection begins when an infected mosquito pierces the human skin it exchanges saliva for human blood and in the saliva can be found numerous sporozoites which are produced in structures called **sporocysts.**

A sporocyst is found near the crop of the insect lying in the connective tissue. It is a prominent swelling which bulges into the haemocoel. Within the sporocyst numerous nuclear divisions occur and vacuoles are formed. Each nucleus becomes surrounded by some cytoplasm to form sickle-shaped sporozoites which arrange themselves around the vacuoles. The sporocyst eventually

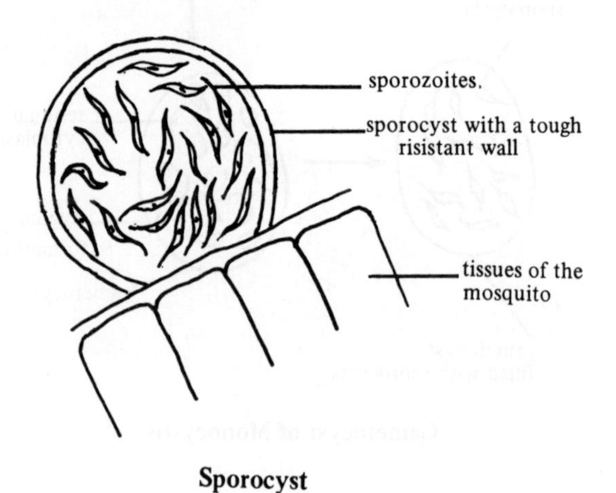

Sporocyst

bursts liberating the sporozoites into the haemocoel and hence to the salivary glands.

When the infected mosquito pierces the skin of the human the parasite-laden saliva is passed into the blood stream and malaria develops.

Statocyst

Statocysts are receptor organs concerned with the detection of changes in the orientation of the medusa of Obelia. Each statocyst is composed of a small fluid-filled ectodermal sac containing a particle of calcium carbonate secreted by a single cell. The sac is composed of epithelium which is sensory on one side and here the cells bear protoplasmic processes which arch over the particle of calcium carbonate. When the medusa is horizontal the statolith hangs downwards and so does not stimulate the sensory processes, but when the medusa is inclined to one side the statolith falls against a sensory cell which becomes stimulated and nerve impulses are transmitted to the cells of the nerve ring. As a result the muscle tails contract more rapidly on the side stimulated so that the bell is brought back to the horizontal position.

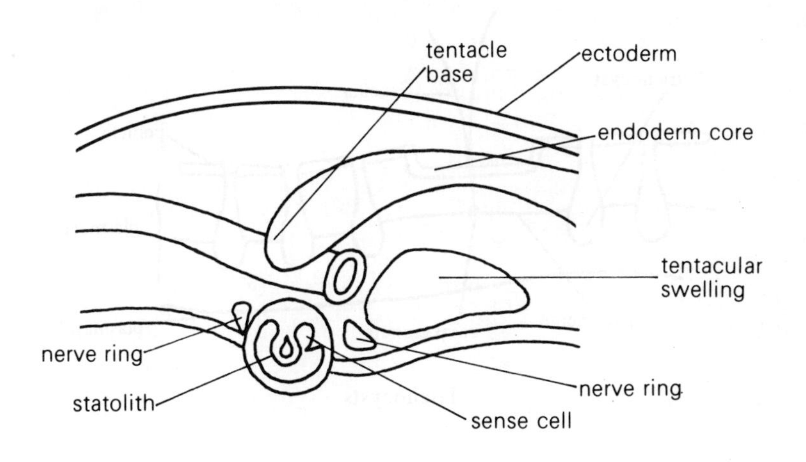

Statolith of Obelia

The statocysts probably encourage swimming movements which keep the medusa in the surface water.

The statocysts occur at the base of each ad-radial tentacle on the sub-umbrella surface just inside the margin of the bell of Obelia.

Trichocyst

Trichocysts are found embedded in the cytoplasm of some Protozoa e.g. Paramecium. They are small spindle-shaped objects which eject long fine threads which may provide the animal with a defence mechanism and may be able to ward off attacks by predators. When the animal is irritated by chemicals trichocysts are ejected. The points are sticky while the rest hardens. They may also be used to anchor the animal while feeding. When they break free the threads are lost but the trichocysts are replenished. Such trichocysts are embedded in the plasmagel and are controlled to some extent by the neuronemes.

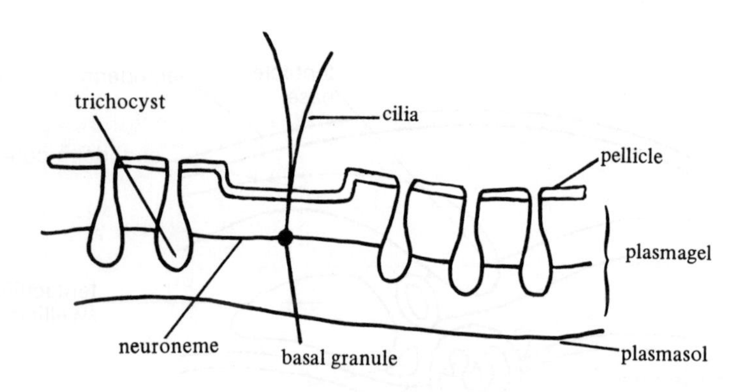

Trichocysts

7. Distinguish between breathing and respiration. Describe the process of oxygen uptake in a named mammal and a named fish.

Respiration in animals e.g. the rat and the dogfish involves two processes, breathing or external respiration and tissue respiration or internal respiration.

The first process, breathing is brought about when oxygen is passed into the blood system of the respiratory organs e.g. the lungs or gills and carbon dioxide is passed out of the blood system into the respiratory organs. This process is a rhythmical movement.

The second process, tissue respiration occurs when the oxygen is carried in the form of oxyhaemoglobin to the tissues where the cells take the oxygen from the blood by diffusion and use it to release energy from the food. Carbon dioxide is produced as a waste product.

The rat is a terrestial mammal and must obtain its oxygen from the surrounding air. The respiratory organs are the lungs which occur within the thoracic cavity. A special respiratory system has developed to channel the air containing oxygen from the outside to the blood system of the lungs. To describe the process of oxygen uptake it is necessary to have some idea of the structure of the respiratory system which is best seen in a diagram (page 24).

When the thorax of the rat is expanded and contracted rhythmically it is said to be breathing. The rat is taking in oxygen (air) into the lungs and breathing out carbon dioxide. This ventilation is affected by movements of the diaphragm and the ribs. The pleural cavity in which the lungs are housed is completely airtight and it is only the interior of the lungs that is actually in contact with the outside air.

When the diaphragm is relaxed it is dome-shaped and the ribs are lowered. To take in air into the lungs the diaphragm muscles must contract and the diaphragm flattens. This movement increases the internal capacity of the thorax. This is also helped by the raising of the ribs. As a result air is drawn into the lungs via the nasal openings, trachea and bronchi and hence gaseous exchange will take place between the air and the blood system of the alvooli.

Once oxygen uptake is complete tissue respiration can occur.

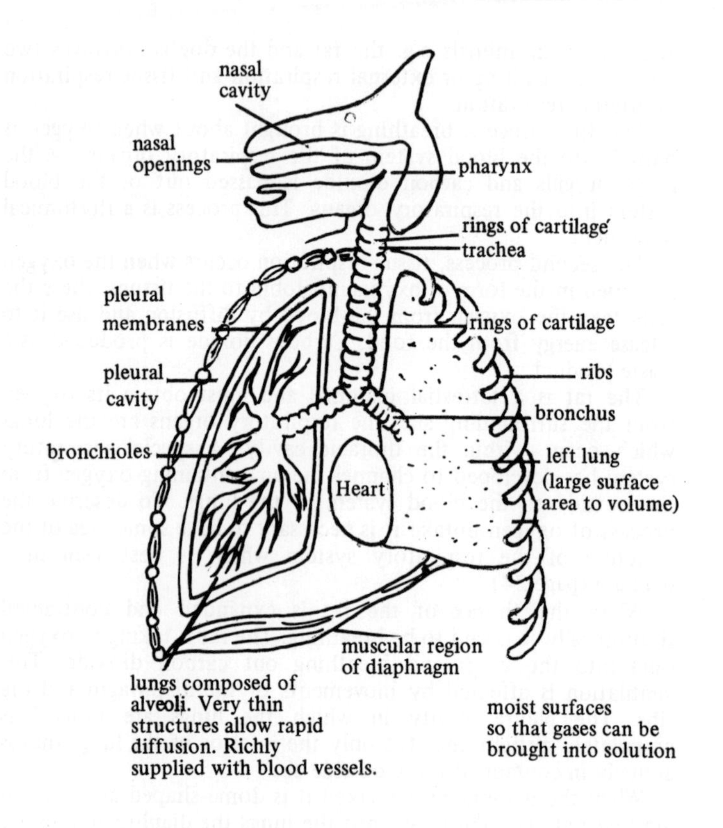

Respiratory system of the rat

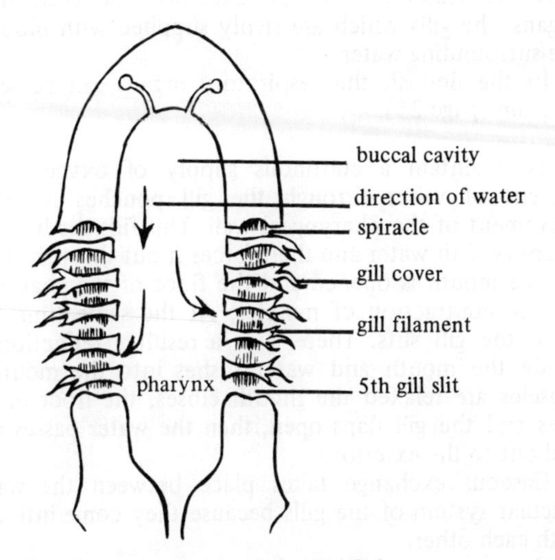

Respiratory system of the dogfish

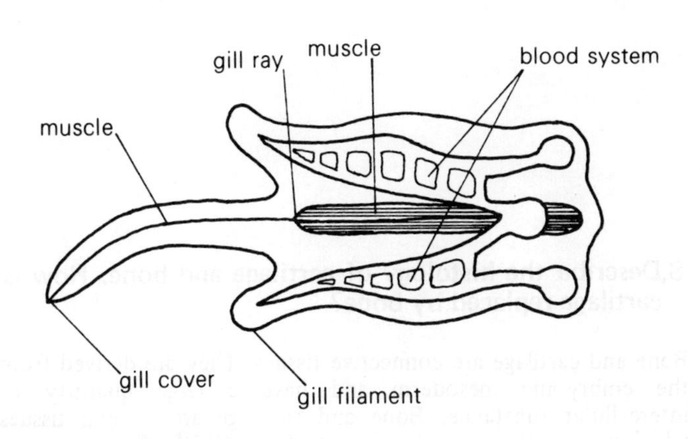

A gill

Fish obtain oxygen for respiration from the water in which they live. Gaseous exchange takes place between the respiratory organs, the gills which are richly supplied with blood vessels and the surrounding water.

In the dogfish the respiratory organs can be seen as in the diagram (page 25).

To maintain a continous supply of oxygen to the gills a current of water through the gill pouches is set up by the movement of the pharyngeal wall. This fills the buccal cavity and pharynx with water and then forces it out through the gill slits.

The mouth is opened and the floor of the pharynx is lowered by the contraction of muscles. At the same time the gill flaps close the gill slits. There is as a result a reduction of pressure inside the mouth and water rushes into the mouth. When the muscles are relaxed the mouth closes, the floor of the pharynx rises and the gill flaps open, then the water passes over the gills and out to the exterior.

Gaseous exchange takes place between the water and the vascular system of the gills because they come into close contact with each other.

Once oxygen uptake is complete tissue respiration can occur.

8. Describe the histology of cartilage and bone. How is cartilage replaced by bone?

Bone and cartilage are connective tissues. They are derived from the embryonic mesoderm and have a large quantity of intercellular substance. Bone and cartilage are skeletal tissues which provide firm areas for the attachment of the tendons of the muscles. They provide support to the body forming a framwork to which soft parts can be fastened. Bones provide a system of levers which with the muscles enable locomotion.

There are three types of cartilage:-

1) Hyaline,
2) Fibro and
3) Elastic.

Hyaline cartilage is a clear bluish coloured glassy substance. It consists of a matrix called chondrin. It is firm and elastic and sometimes fine fibres are present. The embryo skeleton is composed of cartilage, but it is gradually replaced by bone. Cartilage only remains in certain areas such as the xiphisternum and suprascapulars.

Each plate of hyaline cartilage is bounded by a tough fibrous membrane, the perichondrium which bears numerous blood vessels. The chondroblasts that secrete the chondrin lie in small fluid filled spaces called lacunae.

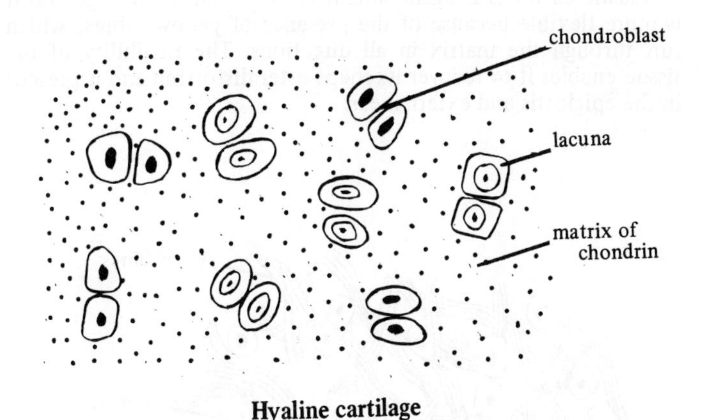

Hyaline cartilage

Fibro cartilage is very similar to hyaline cartilage except that the matrix is packed with bundles of white fibres together with the chondroblasts in the lacunae. Thus the flexibility of the cartilage is combined with the strength of the fibres.

Fibro cartilage is found in the intervertebral discs where the cushioning effect is important.

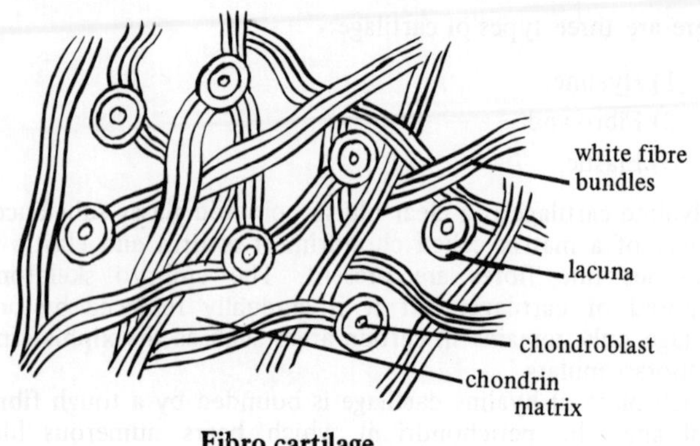

Fibro cartilage

Elastic cartilage is again similar to the hyaline cartilage, but it is more flexible because of the presence of yellow fibres, which run through the matrix in all directions. The flexibility of the tissue enables it to recover its shape after distortion and is present in the epiglottis and external ear.

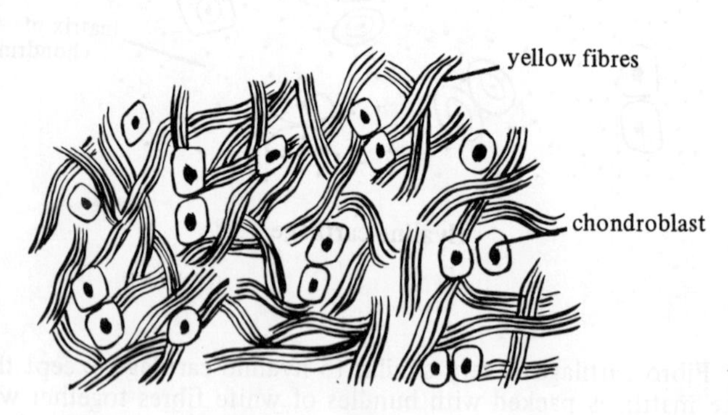

Elastic cartilage

Bone is a connective tissue with a large amount of intercellular substance consisting of mineral salts. A bone is enclosed in a tough sheath of fibrous connective tissue called the periosteum. Beneath the periosteum is a very dense layer of compact bone beneath which is a thicker zone of spongy bone. In the long bones a marrow cavity is present. This is where the blood corpuscles are formed.

The matrix of the bone contains a small amount of white fibres and large quantities of calcium salts. The matrix is perforated by Haversian canals which run parallel with the long axis of the bone. The Haversian canals have an artery and vein to supply the osteoblasts with nourishment. The osteoblasts arrange themselves around the Haversian canals in concentric circles thus forming an Haversian system. Thus forming bone lamallae.

The osteoblasts lie in space called lacunae and these connect with other lacunae. These connections are called canaliculi which contain the fine processes from the osteoblasts so that the whole matrix has a protoplasmic network.

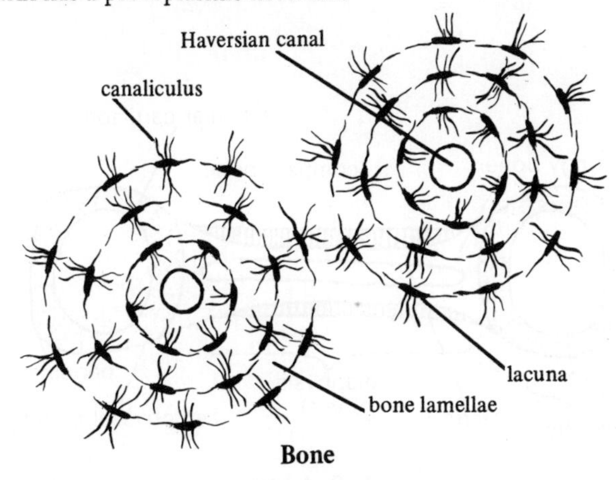

Bone

Ossification of cartilage begins with the cartilage cells multiplying rapidly and arranging themselves in parallel columns in the longitudinal axis of the bone. It begins in the centre and moves outwards. The cartilage becomes known as the periosteum. Osteoblasts proceed to lay down fibres upon the surface of the cartilage. This region becomes calcified and will form compact bone. Next multinucleate cells called osteoclasts erode their way through the sub-periosteal bone into the cartilage making channels for the blood vessels. As a result a series of longitudinal spaces filled with connective tissue, blood vessels and osteoblasts are formed and on the wall of the spaces osteoblasts deposit bone salt. Thus the bone in the shaft is deposited as hollow bars called trabeculae which give it a spongy nature. In long bones the trabeculae are eroded to form the marrow cavity. The spongy bone is surrounded by compact sub-periosteal bone. Later osteoclasts erode the Haversian canals followed by vascular and connective tissue. The two ends of the shaft are still cartilaginous. An ossification centre appears in each epiphysis and as a result an epiphyseal bone is formed. It is separated from the main bone by the epiphyseal cartilage which when growth has finished becomes ossified and further growth stops.

The cartilage gradually becomes replaced by bone which is the same shape as the original cartilaginous bone. However the stresses and strains set up by muscular action often alters the shape of the bone.

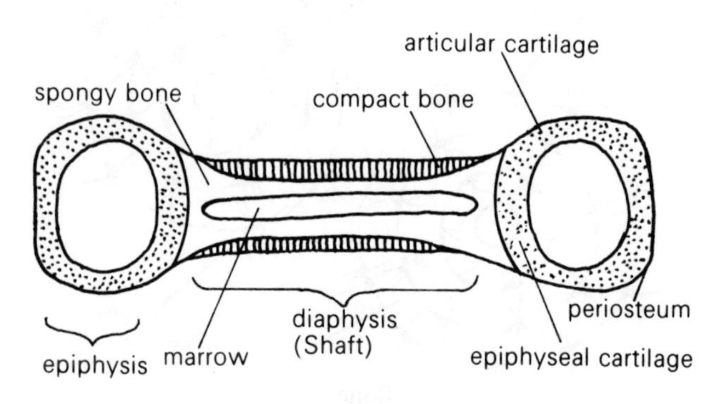

L. S. A bone

9. Give labelled diagrams to show the structure of the pelvic and pectoral girdles of a mammal. What are the functions of these girdles?

The large bony surfaces of the pelvic girdle provide attachment for the limb muscles. The pelvic girdle sustains powerful thrusts of the hind limbs during jumping, while its position parallel to the vertebral column and its fusion to the sacrum ensure that the thrust of the limbs is transmitted to the axis of the body. The elastic symphysis in the female is related to parturition, when the girdle separates to enable the passage of the young down the uterus.

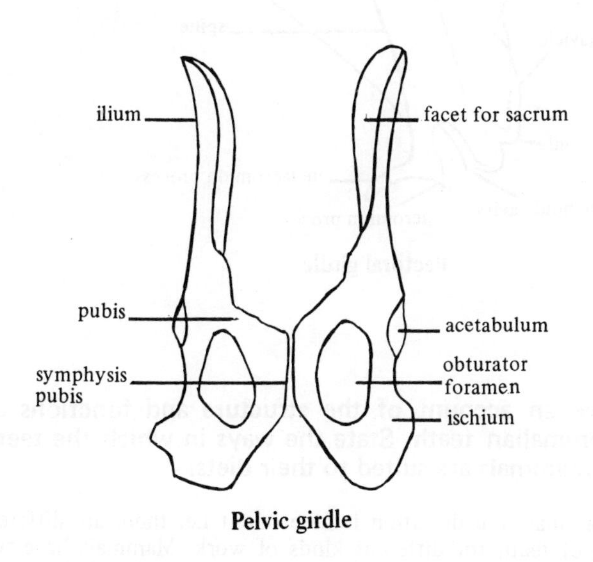

Pelvic girdle

The pectoral girdle reduces the impact of landing and the suppleness of the girdle is important in dispersing the shock. The embedding of the shoulder blade in the muscles of the pectoral region provides sufficient rigidity for the articulation of the forelimb.

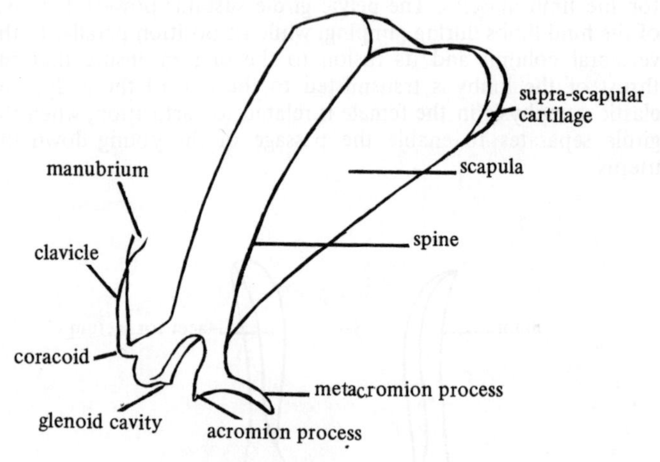

Pectoral girdle.

10. Give an account of the structure and functions of mammalian teeth. State the ways in which the teeth of mammals are suited to their diets.

In mammals the dentition is heterodont i.e. there are different kinds of teeth for different kinds of work. Mammals have two definite sets of teeth and are therefore spoken of as diphyodont. The two sets are the milk set and the permanent set and these two sets allow for the adjustment of the teeth to the enlarging mouth. The teeth are confined to the jaws into which they are set in sockets.

The teeth of all adult mammals have a similar basic structure as can be seen in the diagram.

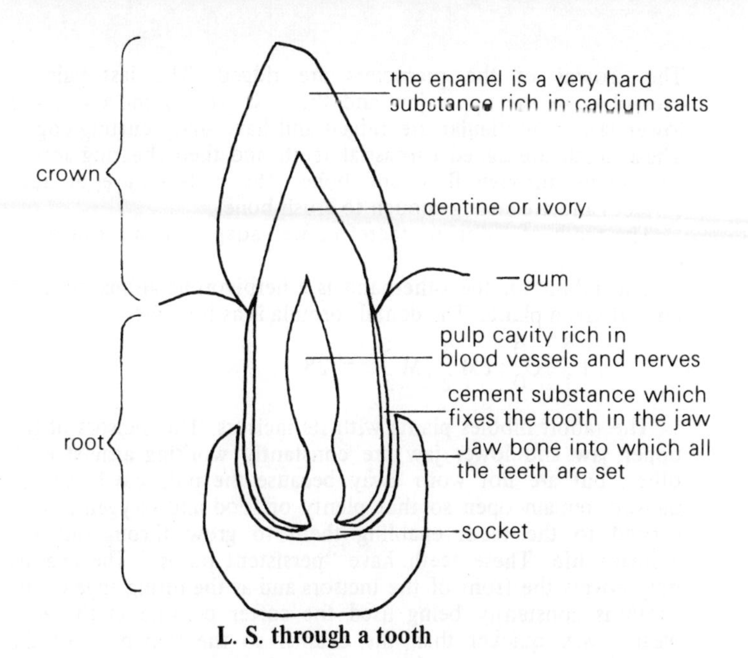

L. S. through a tooth

There is a considerable reduction in the number of teeth in mammals and this results in the shortening of the jaw and a change in food habits.

The types of teeth in mammals are four in number. They are:-

1. Incisors. These are chisel-shaped teeth and are used for biting.
2. Canines. These are long and pointed teeth and are used for digging into flesh, for tearing and fighting.
3. Premolars. These have broad flat surfaces with small cusps or points and are used for grinding.
4. Molars. These are like the premolars, but are a little larger, again they are used for grinding.

The numbers and kinds of teeth found in various animals are usually expressed by a dental formula which represents half the upper and lower jaws.

In the dog, the dental formula is:-

$$I\frac{3}{3}; C\frac{1}{1}; PM\frac{4}{4}; M\frac{2}{3}. = 42$$

The dog is a carnivorous animal and is naturally adapted for catching and feeding on other animals. Their incisors are used for biting and the curving canines are well developed for striking at their prey. The premolars and molars are not used for grinding.

The crowns of the premolars are ridged. The last pair of premolars in the upper jaw and the first pair of molars in the lower jaw in particular are ridged and have sharp cutting edges. These teeth are called carnassial teeth and their shearing action cuts easily through flesh and bone. The molars have broader surfaces and are strong enough to crush bone.

The teeth of a dog are therefore well adapted to a carnivorous diet.

The rabbit on the otherhand is a herbivorous animal feeding on soft green plants. The dental formula is as follows:-

$$I\frac{2}{1}; C\frac{0}{0}; PM\frac{3}{2}; M\frac{3}{3}. = 28$$

The rabbit nibbles plants with its incisors. The incisors of the upper jaw and lower jaw are constantly working against each other, but are not worn away because the pulp cavity of the incisors remain open so that plenty of food and oxygen can be carried to the teeth enabling them to grow throughout the animal's life. These teeth have "persistent pulps." The enamel only covers the front of the incisors and as the biting edge of the tooth is constantly being used the softer dentine at the edge wears away quicker than the enamel so the thin plate of the enamel forms a cutting edge.

These teeth would continue to grow and would eventually interfere with the feeding habits of the rabbit if they were not worn away by gnawing. The rabbit has no canines and the space in the jaw normally occupied by these teeth is called the distema. It is overlapped by fur which prevents any food in the mouth from escaping through the gap. The furry lips also help to draw food in and help the tongue to push the food to the back of the mouth. The premolars and molars have flat surfaces crossed by bars of hard enamel. These teeth are excellent grinding structures and completely break down food into fine particles before being swallowed by the rabbit.

Man is an omnivorous animal feeding on both plant and animal materials. The dental formula for man is:-

$$I\frac{2}{2}; C\frac{1}{1}; PM\frac{2}{2}; M\frac{3}{3}. = 32$$

As in most mammals the incisors are used for biting food and are chisel-shaped with sharp edges. They are not used for gnawing and therefore do not have "persistent pulps" as in the rabbit. The canines are not long and pointed and are not used for catching prey, as in the dog. They remain no longer than the incisors, although they are deep rooted. The premolars and molars have broad flattened surfaces with hard enamel ridges crossing them and are typical grinding teeth.

Thus although the dog, the rabbit and man are all mammals they have different feeding habits and as a result their teeth have become adapted to their various diets.

11. Describe how carbohydrates are digested and utilized by the mammalian body.

Digestion is the breaking down of food materials into smaller particles which can be readily absorbed through the gut walls of an animal and hence utilized by the cells of the body.

The digestive process of the carbohydrates begin in the mouth. The food is bitten off by the incisors and passed to the back of the buccal cavity by the tongue and ground up by the molars. The salivary glands secrete a fluid called saliva which mixes with the masticated food. This breaking down of the food exposes a larger surface area to the action of the enzyme ptyalin. Ptyalin begins the digestion of any cooked starch and glycogen to maltose and dextrins. The ptyalin works in an alkaline medium in the mouth. Mucus is secreted with the saliva which lubricates the buccal cavity and food. A bolus of food is passed to the pharynx and swallowed and then the bolus passes down the oesophagus by a peristaltic movement. The muscular movements cause a constriction of the oesophagus behind the bolus and a relaxation in front of it. The food bolus passes into the stomach.

The stomach is a muscular bag in which food is stored for up to two hours in man. The enzymes of the stomach which are secreted by the gastric walls do not digest carbohydrates, but the hydrochloric acid may hydrolize sucrose to glucose and fructose.

The food becomes a thick creamy paste in the stomach and it is called chyme. This chyme leaves the stomach via the pyloric sphincter by the muscular activity of the stomach. Once the chyme enters the duodenum its acidic nature stimulates the formation of the hormone secretin which once in the blood passes to the pancreas which is stimulated to secrete its pancreatic juices and similarly an hormonal response of the duodenal walls stimulates these cells to produce their digestive enzymes.

The acidic nature of the chyme is neutralized by the presence of bile. Bile is manufactured in the liver and stored in the gall bladder. The salt present neutralizes the chyme. There are no digestive enzymes in bile.

The pancreatic juice however contains several enzymes, one of which is involved in the digestion of carbohydrates. Amylase finishes the work of the ptyalin, therefore completing the digestion of starch and glycogen to maltose. The pancreatic juice is alkaline in man due to the presence of sodium bicarbonate. The food continues to move along the duodenum where more digestive juices, the succus entericus comes into contact with it. The succus entericus is a watery secretion containing several enzymes, three of which are carbohydrases, invertase, maltase and lactase. These three enzymes complete the digestion of carbohydrates in the body. The invertase converts sucrose into glucose and fructose; maltase converts maltose into glucose; lactase converts lactose into glucose and galactose.

Thus the chyme is continually being hydrolysed by the enzymes and the chyme becomes even more watery and is called chyle.

The carbohydrates have been digested to the most soluble form and are now suitable for absorption. This takes place in the small intestine. The internal surface of the small intestine is increased by the presence of numerous villi. Each villus is supplied by a vascular system so that the absorbed materials may be transported rapidly from the gut.

The carbohydrates are absorbed as monosaccharides. All carbohydrates not required for respiratory purposes are degraded to pyruvic acid which is decarboxylated to "active acetate" with co-enzyme A. The "active acetate" is synthesised into fatty acids and then to fats and stored in the fats depots e.g. under the skin and around the kidneys.

Carbohydrates are transferred to the liver and built up into glycogen and stored in the liver cells. Other stores of glycogen are in the muscles. Glycogen is broken down to glucose in the liver and transferred via the blood to the muscles to be resynthesised into glycogen until required for respiration.

The amount of glucose in the blood of a mammal is regulated by the liver. A rise in the blood sugar content causes deposition of glycogen in the tissues and liver; a fall causes glycogen to be converted back into free glucose in the blood.

Carbohydrates play an important part in the life of animals but before they can be utilized they must be digested into their simplest form.

12. Describe the structure and functions of the vertebrate liver.

The liver is the largest gland in the body. It is a reddish purple organ lying on the posterior side of the diaphragm and partly covering the stomach. The whole organ is enclosed in a double sheet of periosteum. The liver is made up of five lobes; two main lobes sub-divided into smaller lobes. Embedded in the right side lobe is the gall bladder from which a narrow duct passes to open in the duodenum. The gall bladder stores a green alkaline liquid called bile.

The liver is composed of liver cells with a prominent nucleus with large quantities of glycogen and fat droplets. These cells form a continuous mass of cells tunnelled by lacunae through which run the blood capillaries or sinusoids. The sinusoids are derived from branches of the hepatic portal vein. The sinusoids receive blood from the branches of the hepatic artery. All the blood vessels can constrict their lumen and thus vary the blood supply to the liver cells.

The liver cells appear to be grouped in polygonal lobules bounded by portal canals. The portal canals are large organs which carry the hepatic portal vein, hepatic artery, lymph vessel and branches of the bile duct.

The polygonal lobules are only temporary for they are dependent on the pressure gradients in the vessels crossing the liver. Around each liver cell are minute bile canaliculi into which bile is secreted. These canaliculi join together and gradually join up to form the bile duct.

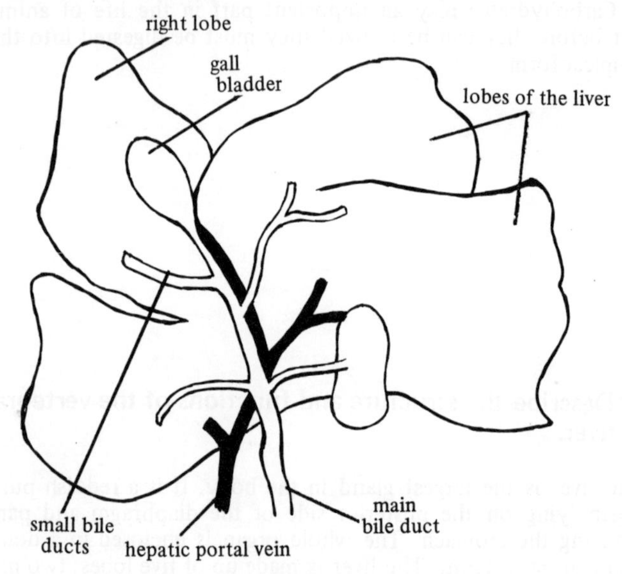

The liver

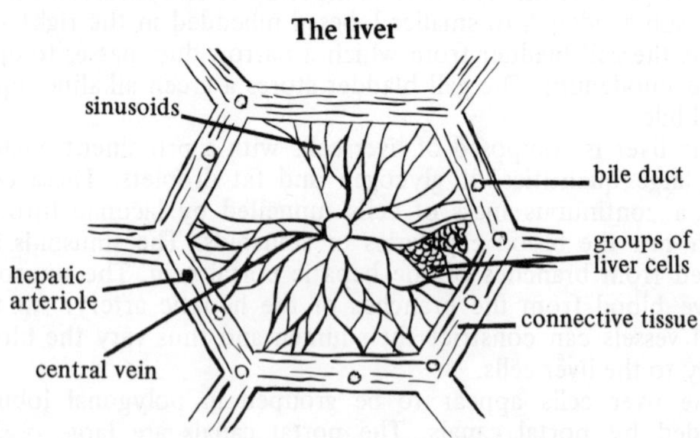

Histological appearance of the liver

The liver plays a very important part in the life of an animal. Amino acids absorbed from the gut are taken to the liver by the hepatic portal system. Amino acids are required for the building up of the proteins necessary for growth and repair. Amino acids are therefore distributed around the body from the liver via the blood system. Amino acids are not stored in the body and any excess must be metabolised as soon as possible in the liver. A process of deamination occurs which involves the removal of nitrogen and the production of urea which is eliminated by an excretory process.

Secretion of bile is caused by a stimulation from the vagus nerve. The bile is an alkaline green liquid rich in sodium salts. It also contains two pigments bilirubin and biliverdin. The bile pigments are the products of the breakdown of haemoglobin which takes place in the liver. They contain no enzymes and take no part in digestion. Bile also excretes excess cholesterol as well as the pigments.

The bile salts lower the surface tension of the fats enabling them to form an emulsion and enable lipase to work in the duodenum. The alkaline nature of bile neutralises the acidity of the chyme.

Carbohydrate metabolism is centred around the glycogen content of the liver. Glycogen can be built up in the liver from absorbed carbohydrates and noncarbohydrates e.g. amino acids. Hexose sugars can be changed into glycogen, but when the glycogen is broken down it always forms glucose. Thus the liver is the store house for glycogen.

The liver regulates the blood sugar content and a rise in the blood sugar ratio causes glycogen to be deposited in the tissues of the liver. A fall causes the conversion of glycogen in the liver to glucose, and free glucose in the blood. This occurs through a number of hormones. Insulin plays an important part, for insufficient insulin produced causes diabetes and a high blood sugar content.

The amount of glycogen stored in the liver is limited. The carbohydrate intake not converted into glycogen is converted into fat and stored in the fat depots of the body e.g. beneath the skin and around the kidney.

13. Give an account of the structure and function of the heart of a named mammal.

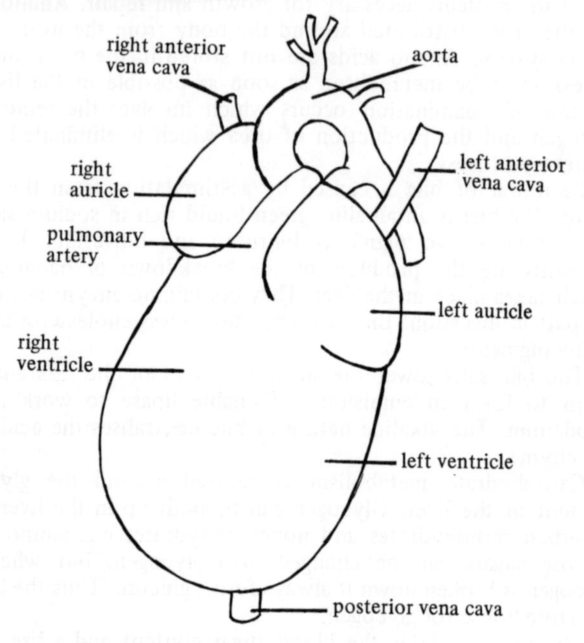

Ventral view of the rabbit's heart

The heart of the rabbit lies ventrally in the thoracic cavity with its apex slightly tilted to the left. It is covered by coelomic epithelium, the pericardium. The heart consists of four chambers, the right and left auricles and the right and left ventricles. All four chambers have muscular walls which differ in thickness. The auricles are thin walled chambers and are situated anteriorly, while posteriorly lie two much larger thick walled ventricles.

Three main blood vessels enter the right auricle, two anterior venae cavae and one posterior vena cava. While entering the left auricle are two pulmonary veins, one from each lung.

The right auricle communicates with the right ventricle through an opening called the auriculo-ventricular opening and the left auricle communicates with the left ventricle.

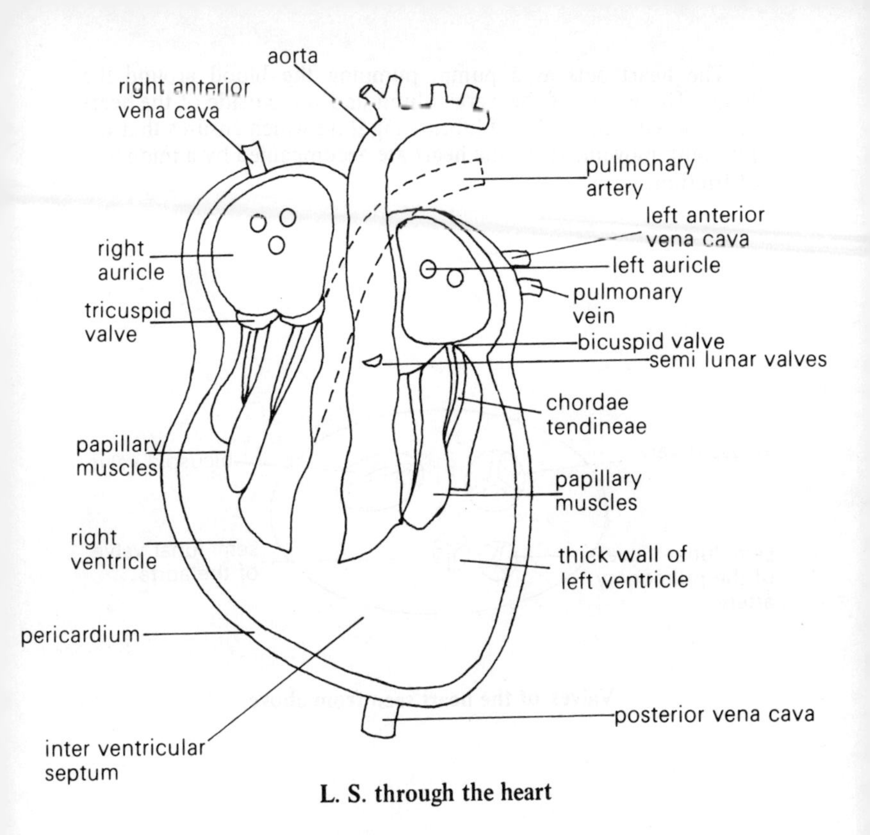

L. S. through the heart

The opening between the right auricle and right ventricle can be closed by three fibrous flaps projecting from the walls, this is the tri-cuspid valve. The free edges are anchored to the inside of the wall of the right ventricle by the chordae tendineae. The opening between the left auricle and the left ventricle has two similar flaps the bi-cuspid valve or mitral valve joined to the inside of the left ventricle by more chordae tendineae.

The left side of the heart is completely cut off from the right side by the inter-auricular septum between the auricles and the inter-ventricular septum between the ventricles

An artery opens from each of the ventricles, the pulmonary artery from the right and the aorta from the left. At the base of each artery are three semi-lunar valves which prevents the back flow of blood into the heart.

The heart acts as a pump, pumping the blood around the body. The inside of the pericardium and the outside of the heart are covered with a film of lubricating fluid which ensures that the pumping movements of the heart are accompanied by a minimum of friction.

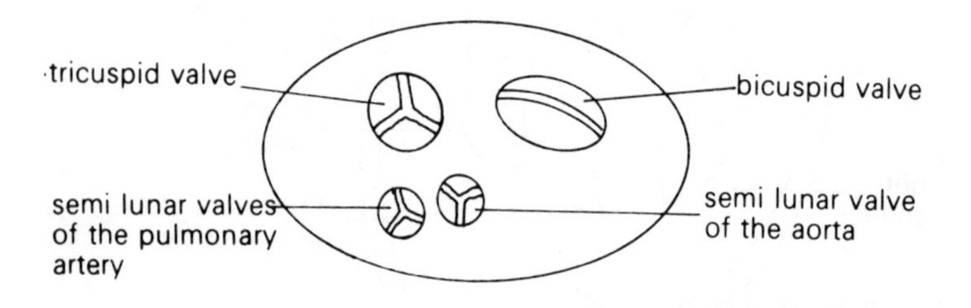

Valves of the heart seen from above

The auricles fill with blood until the pressure closes the valved entrances of the venae cavae and the pulmonary veins. The auricles once filled with blood contract and push the blood through the tri-cuspid and bi-cuspid valves and into the ventricles. Now the ventricular pressure rises and the tri-cuspid and bi-cuspid valves are closed. The ventricles contract opening the valves of the aorta and pulmonary artery and the blood is forced out of the heart. Thus a double circulation of blood through the heart is accomplished.

14. Draw and label a diagram of the mammalian ear. Give an account of its functions.

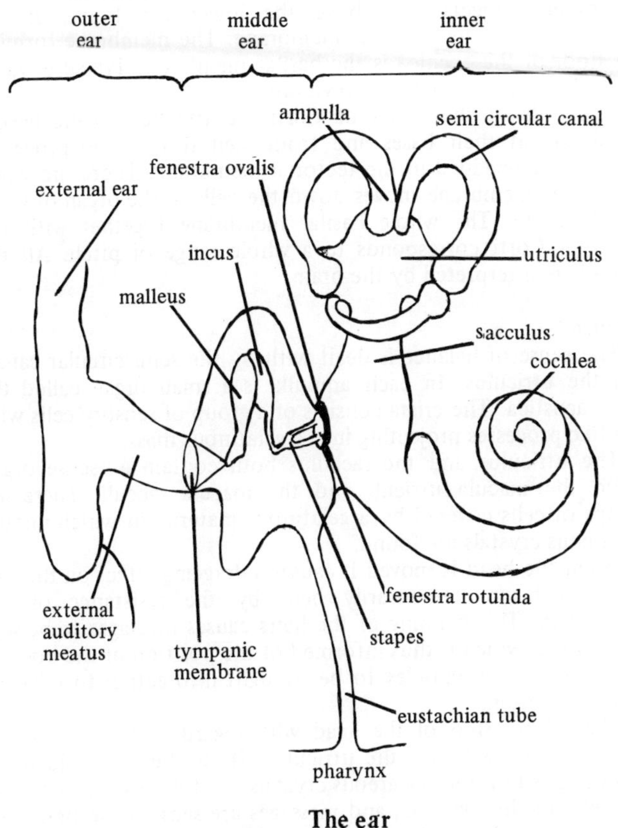

The ear

The ear is concerned with the sense of hearing and with balance.

Hearing

Air pressure waves strike the tympanic membrane which vibrates and causes the ossicles to move. The stapes vibrates on the fenestra ovalis which in turn sets pressure waves in the perilymph surrounding the cochlea. Inside the cochlea the endolymph vibrates which agitates special cells at the base of the cochlea. These special cells comprise the organ of Corti and when disturbed cause impulses to be sent along the auditory nerve from the branches of this nerve to the nerve branches of these cells.

The organ of Corti has the following structure. A fine membrane projects out above the lower membrane of the cochlea, this is the tectorial membrane. The membrane forming the floor of the cochlea is the basilar membrane. Pressure waves cause the basilar membrane to vibrate.

The cells of the organ of Corti are attached to the basilar membrane at their bases and from their tips extend processes which are embedded in the tectorial membrane. Therefore when the basilar membrane moves down the cells of the organ of Corti are distorted. The whole basilar membrane together with the organ of Corti corresponds to a whole range of pitch. All the stimuli are interpreted by the brain.

Balance

The sense of balance is dealt with by the semi circular canals and the utriculus. In each ampulla is a small organ called the crista acustica. The crista consists of a group of sensory cells with hair like processes projecting into a gelatinous mass.

The utriculus and the sacculus both contain a sense organ called the macula utriculi and the macula sacculi. These are groups of cells covered by a gelatinous material in which minute calcareous crystals are found.

When the head is moved it causes a dragging effect on the hair cells which are temporarily bent by the resistance of the endolymph. This bending of the hairs causes messages to be sent to the brain which is thus informed of the position of the body in space and causes muscles to be brought into action to take the necessary action.

The relationship of the head with regard to the rest of the body is dealt with by the utriculus. If the head is held in an unusual position the calcareous crystals or otoliths exert a pull on the cells of the maculae, and messages are sent to the brain and the muscles right the body. It is the effect of gravity on the sense cells of the utriculus and sacculus which tells an animal whether it is the right way up and therefore enable it to take the necessary steps to right its position.

Thus the semi-circular canals are concerned with the movements of the head, while the utriculus and sacculus with the action of gravity and the position of the body as a whole.

15. By means of fully labelled diagrams only compare the excretory and reproductive systems of a named male and female mammal.

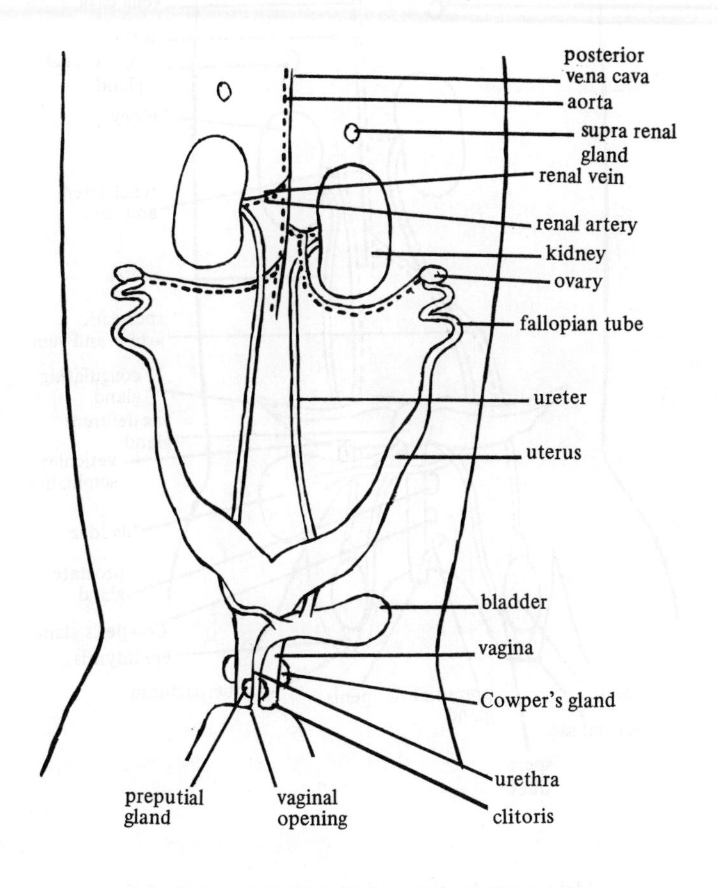

Female excretory and reproductive systems of the rat

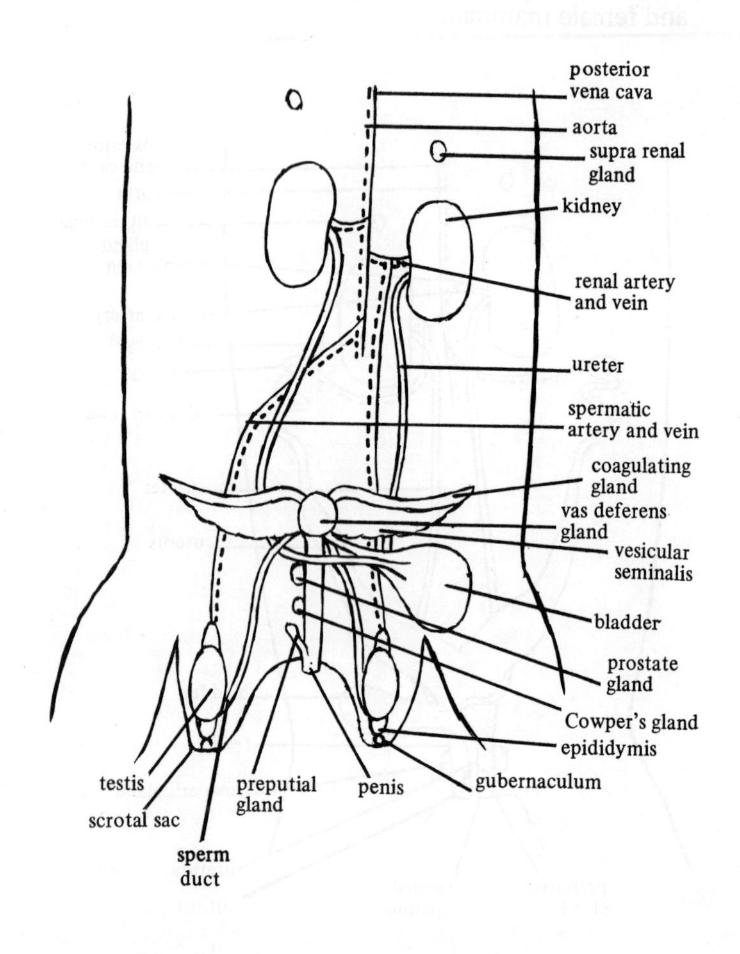

Male reproductive and excretory systems of the rat

16. Describe the gross and minute structure of the mammalian kidney. What are the functions of the various parts.

The kidneys are a pair of dark red bean-shaped organs attached to the dorsal wall of the abdominal cavity. They are on either side of the lumbar vertebrae. Each has a renal artery and a renal vein. Attached to the concave side of each kidney is a tube called a ureter which leads to the bladder.

The kidney is composed of two parts. The outer region is called the cortex which has a granular appearance and the inner region the medulla which is striated. The whole kidney is contained within a tough connective tissue called the capsule.

The cortex contains numerous small thin walled structures

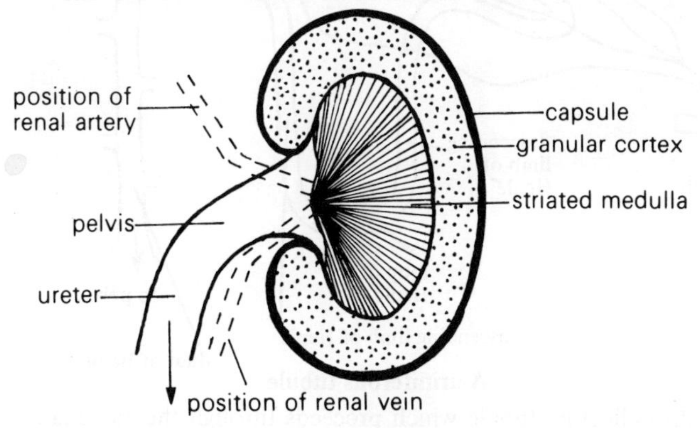

The kidney

called tubules. Each tubule ends blindly as a small rounded funnel called Bowman's capsules. Each Bowman's capsule contains a knot of capillaries, the glomerulus. Leading from each capsule is a fine tubule in which three sections can be distinguished. A convoluted tube next to the capsule winds in the cortex and then a descending portion which is called the descending limb of Henlé which passes into the medulla. Then the tube passes back at the U-shaped loop of Henlé most of which projects into the medulla and then the ascending limb of Henlé returns to the cortex where it has a second convolution which with similar tubules enters a

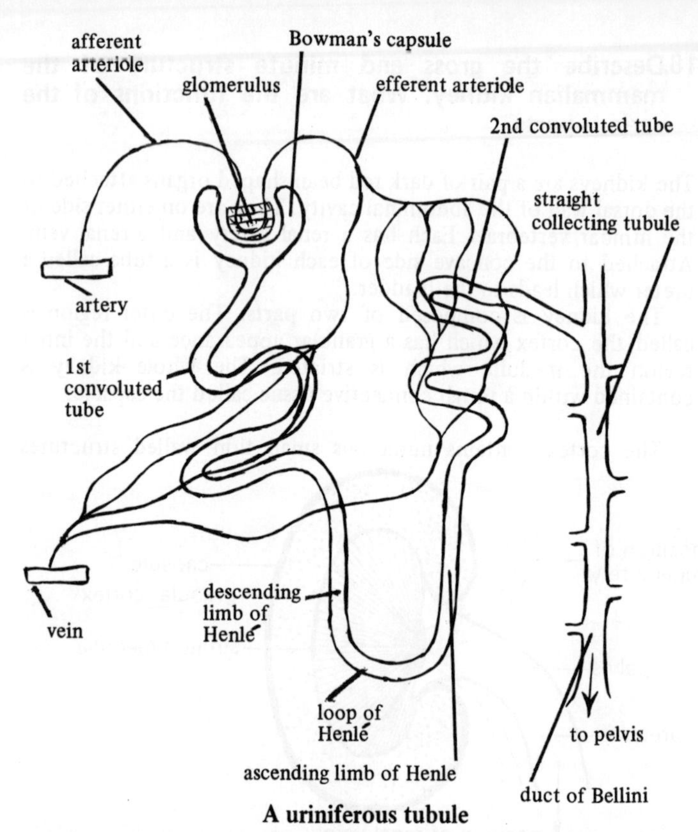

A uriniferous tubule

straight collecting tubule which proceeds through the medulla to open at the apex of the pyramid in the duct of Bellini and hence into the pelvis from where the ureter leads to the bladder.

A high pressure is set up in the blood entering the glomerulus and as a result water and simple solutes pass into the capsule by a process called ultra-filtration. This liquid contains urea which was synthesised in the liver, glucose, mineral salts and water. All but the urea are useful to the body and so must be reabsorbed.

The function of the uriniferous tubule is to alter the composition of this liquid and in the convoluted portions of the tubule the valuable solutes are reabsorbed. The reabsorption process is considered to be an active process and not a passive diffusion process. The tubule is rigid which suggests an energy output.

The weakened filtrate continues along the loop of Henlé and

at the distal parts of the tubule further reabsorption takes place. Even further along the tubule the character of the wall changes for it becomes more or less waterproof and prevents reabsorption. Here the filtrate has the same composition as urine. The urine passes through the collecting tubule to the pelvis then out of the kidneys into the ureter and to the bladder.

The kidneys act as osmo-regulators of the water content of the body reabsorbing more or less from the glomerular filtrate according to the needs of the body. They also serve to get rid of the nitrogenous waste materials of the body.

17. Write a short account of the functions of each of the following

a) thyroid,

b) pituitary,

c) adrenals,

d) Islets of Langerhan.

The thyroid gland produces the hormone thyroxin which is rich in organically combined iodine. It is necessary for the normal growth and development of the animal. The gland will not work correctly in the absence of the thyrotropic hormone from the pituitary.

Underactivity of the gland in a child results in cretinism which is characterised by small stature and mental backwardness. In adults it causes myxoedema marked by a slow mental and physical activity and puffiness of the hands and face. Both cases are treated by giving thyroid extract orally.

Deficiency is often caused by a lack of iodine in the diet. The gland may enlarge producing a swollen neck condition called a goitre. This can be remedied by administering iodine compounds in table salt.

In eels and tadpoles absence of the hormone prevents metamorphosis.

The pituitary gland is composed of two lobes, the anterior and the posterior. Each lobe secretes hormones which have a considerable affect on the animal body.

The anterior lobe produces a growth hormone which affects protein metabolism. Excess of the hormone before maturity results in the overgrowth of the long bones and therefore giants are formed. Excess of the hormone after maturity does not affect skeletal size, but malformations of the hands and feet and face may occur.

Deficiency of the hormone before maturity results in dwarfism.

Hormones are produced which influence the activity of the thyroid gland, the adrenal glands, the islets of Langerhan and the gonads.

The posterior lobe also secretes important hormones. Vasopressin is secreted which stimulates the muscle fibres of the arterioles. It acts independently of the nervous system and constricts the arterioles causing a rise in the blood pressure.

Other secretions affect the muscles of the uterus and the control of water absorption in the kidney tubules. Carbohydrate metabolism is affected by secretions from the posterior lobe of the pituitary.

The pituitary gland as a whole has a marked affect on the development and activity of the mammalian body.

The adrenals like the pituitary has two parts, they are called in the adrenals the cortex and the medulla.

The cortex secretes the hormone cortin which controls metabolism. It is involved in controlling the ionic distribution of sodium and potassium. Without cortin the animal dies. It is also involved in the metabolism of carbohydrate. Deficiency of the cortical hormone results in Addison's disease.

The medulla secretes adrenalin which causes the constriction of the blood vessels in the skin and the dilation of those in the muscles. It raises the heart beat and the sugar content of the blood so that the muscles receive an additional supply of oxygen and sugar. Fright and anger causes an increase in the adrenalin content of the blood through the medium of the nervous sytem, so that the body is prepared for any emergency.

The Islets of Langerhan produce the hormone insulin. This enables the liver and the muscles to store glycogen and to increase the ability of the muscles to oxidise glucose.

Deficiency results in the disease diabetes in which sugar is excreted in the urine. This can be remedied by insulin injections.

18. Give an account of the way in which oxygen is transferred from the outside environment to the tissues of the body in the earthworm, cockroach and a flowering plant.

Most plants and animals require oxygen for the liberation of energy during respiration and it is this energy that is required for all the necessary life processes to take place.

The animals and plants obtain their oxygen from the surrounding environment. The earthworm is a terrestial animal and obtains the necessary oxygen by diffusion through its skin.

Section through the skin of the earthworm

The cuticle is thin and is kept moist by secretions from the epidermal mucous glands and from fluid from the dorsal pores. The epidermis of the earthworm is richly supplied by a looped capillary blood system which comes very near the surface. The blood contains the respiratory pigment haemoglobin which enters into chemical combination with the oxygen to form oxy-haemoglobin. The ratio of surface area to volume remains constant because of the elongated shape of the body. The oxhyaemoglobin circulates with the blood and releases the oxygen to the cells where there is a low oxygen tension. It must be emphasised that the blood does not come in direct contact with the tissue cells and that the oxygen is passed on through a tissue fluid intermediary.

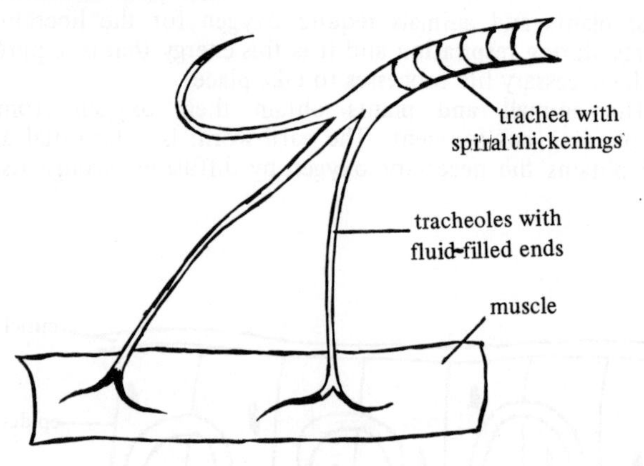

Cockroach tracheole

The cockroach has a branched network of tubes which form a tracheal system and enables oxygen to penetrate all parts of the body. Each trachea consists of squamous epithelium which secretes a cuticular lining and since the linings are in contact with the exoskeleton they are shed during each ecdysis. All the trachea are strengthened by a spiral band of material which prevents them from collapsing, but allowing them to bend when parts of the insect moves. The trachea diminish in size as they branch and they finally end in minute intercellular tracheoles without linings. The tracheoles are not shed during ecdysis, but grow to meet the new physiological demands of the insect. The tissues of the body are penetrated by the fine air vessels and this ensures that no cell is far from a supply of oxygen.

Air enters the body by means of ten pairs of spiracles which are narrow oval slits in soft round patches of cuticle. The spiracles are controlled by valves operated by special muscles. Internally they lead into air sacs from which the tracheal trunks branch.

There is a continuous uptake of oxygen from the tracheoles by the cells and an oxygen gradient is set up between them and the spiracles and oxygen will diffuse into the tracheae from outside. When at rest oxygen can reach the tissues by this means. When active the insect has special breathing movements of the abdomen. If the spiracles are open contraction will force the air out, if they are closed then contraction will force the air deeply into the tracheoles. The tracheoles are partly filled with fluid, but when the metabolic rate is high the fluid is withdrawn from the tracheoles due to a rise in the osmotic pressure of the cells and thus air comes into closer contact with the tissues and they obtain the necessary supply of oxygen for respiration.

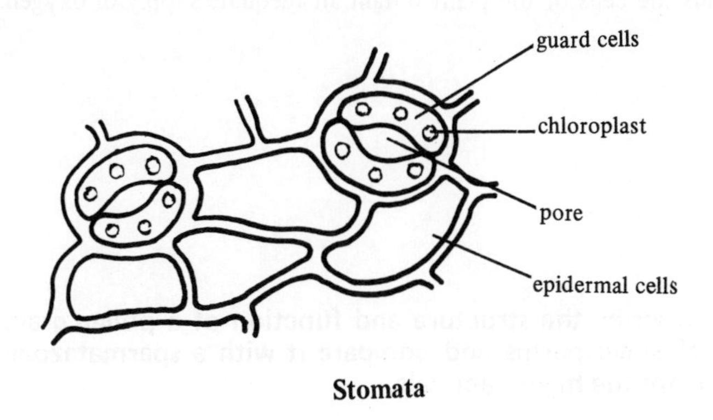

Stomata

Plants such as Helianthus annus rely on a ventilating system of channels and air spaces between the tissues. This internal ventilating system is in communication with special apertures called stomata. Through these gaseous exchange takes place. It is possible that if an adequate water supply is not present a rapid

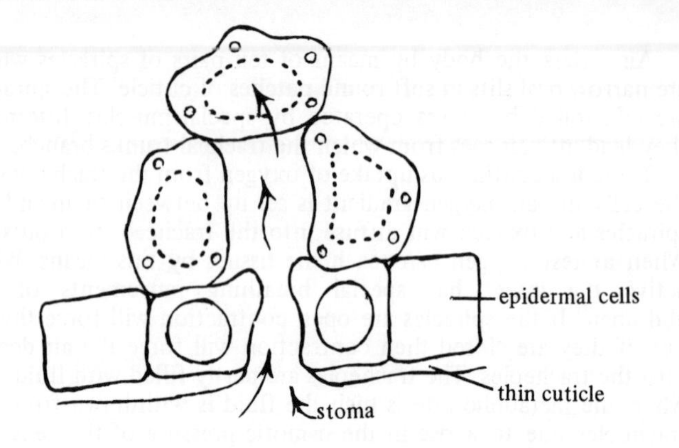

The passage of gas through a stoma

loss of water and the drying out of tissues could result and so some plants have special modifications to prevent this occurring. Thus the cells of the plant obtain an adequate supply of oxygen.

19. Describe the structure and function of a pollen grain of angiosperms and compare it with a spermatazoan from the higher animals.

Pollen grains are produced in the pollen sacs of anthers. They are produced from sporogenous cells called pollen mother cells which undergo meiotic division to eventually form four haploid pollen grains.

Each pollen grain has a two layered wall, an outer exine and an inner intine wall. The exine is first formed and is rarely complete all over the grain. Cracks and fissures are left through which the pollen tube breaks when the grain germinates. The exine may be very thick and spiny, and in some species sticky.

The intine is much more delicate and composed of cellulose and pectic substances. Each pollen grain contains two nuclei, the generative nucleus from which two male gametes will be derived and the pollen tube nucleus.

The pollen grain can not produce a new individual alone for it must combine with an egg of the same type. For this to occur pollination must take place. This is the transference of pollen from the anther of the stamen to the stigma of the carpel. Once pollination is affected the pollen grain germinates and a pollen tube breaks through the outer covering of the pollen grain. This pollen tube results from the expansion of the intine through the cracks in the exine.

The pollen tube carrying the pollen tube nucleus penetrates the tissues of the stigma, style, carpellary wall and nucellus of the ovule, to make contact with the embryo sac. It is through this tube that the male gametes pass. The gametes are non motile and are conveyed in the cytoplasm at the tip of the advancing tube. This is achieved by the secretion of enzymes into the carpellary tissue. When the embryo sac has been penetrated the tip of the pollen tube opens and the gametes enter the embryo sac. The first passes between the synergidae and fuses with the egg to form the zygotic nucleus. The second passes the diploid fusion nucleus and migrates further into the embryo sac until it reaches the polar nuclei and all three nuclei fuse together. These acts of nuclear fusion complete fertilisation and the function of the pollen grain is ended.

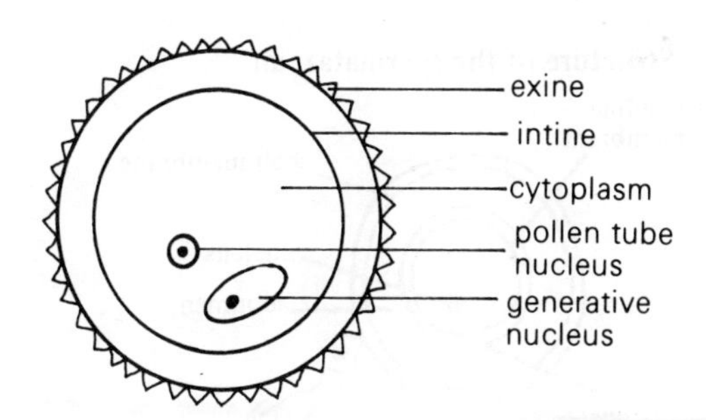

Structure of the pollen grain

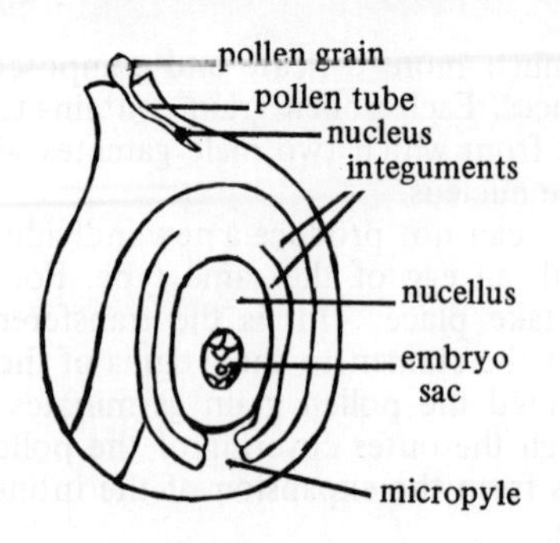

Function of the pollen grain

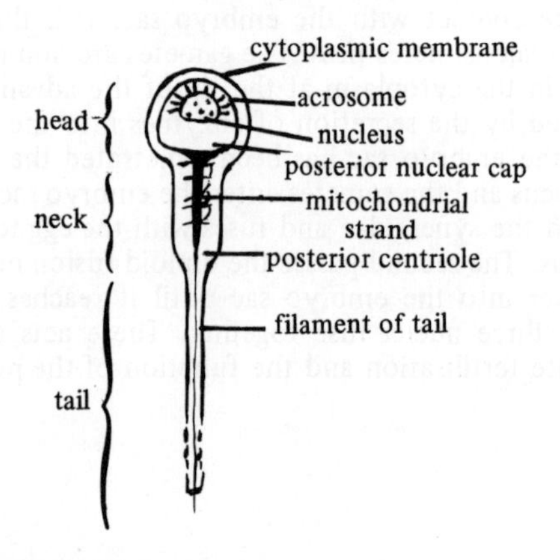

Structure of the spermatazoan

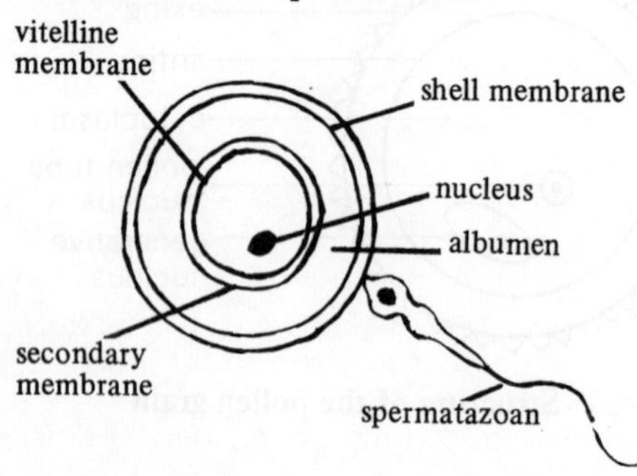

Function of the spermatazoan

Both pollen grains and spermatozoa are produced in large numbers. The spermatazoa are also produced in special sex organs in this case the testis. The spermatazoa arise from meiotic divisions of special cells called spermatocytes. Each spermatocyte gives rise to four special spermatids each of which will develop into a spermatazoan, the structure of which can be seen in the diagram.

A spermatazoan is motile but before fertilisation can take place it must be placed near to the egg and this occurs during mating. Once near the egg their motility enables them to reach the outer covering of the egg. The liquid in which the sperms are deposited contain activators that stimulate the sperms locomotion. Each sperm can only live for as long as its food reserves hold out, about 24 hours in the higher animals.

The sperm penetrates the outer covering of the egg by means of its acrosome. It casts off its tail and the head and middle piece are drawn deeply into the egg cytoplasm. The result is a mixing of the haploid nuclear material to produce a diploid nucleus of the zygote.

Both pollen grains and spermatazoa are produced in large numbers in individuals that are sexually mature. The zygotes produced as a result of fertilisation in both animal and plant is unique among cells as their potentialities are enormous for from them are derived all the cells capable of producing a new organism.

20. What is a gamete? Describe the structure and behaviour of gametes of Chlamydomonas, Spirogyra and Fucus.

A gamete is a reproductive cell whose nucleus must fuse with that of another gamete to give rise to a new organism.

In colder weather a mature Chlamydomonas may divide into a large number of smaller individuals. All the individuals produced are of equal size and similar in structure to the parent except that they have no cell wall and are much smaller than the parent. These individuals are biflagellate gametes.

When these gametes are released from the parent they seek out a gamete from another cell and the two come in contact at their anterior ends. The union of each pair of gametes is called conjugation. The flagella are withdrawn and the two gametes merge into one. The cell formed is called a zygote and develops a thick wall around itself to form a zygospore. The zygospore is able to survive the cold conditions by lying dormant in the mud. Later the contents divide meiotically to form four protoplasts, which change into four biflagellate cells. All the Chlamydomonas cells are haploid, the zygote only being diploid. The cells are released when the thick outer wall of the zygospore is shed and eventually the cells mature into the adult vegetative structure.

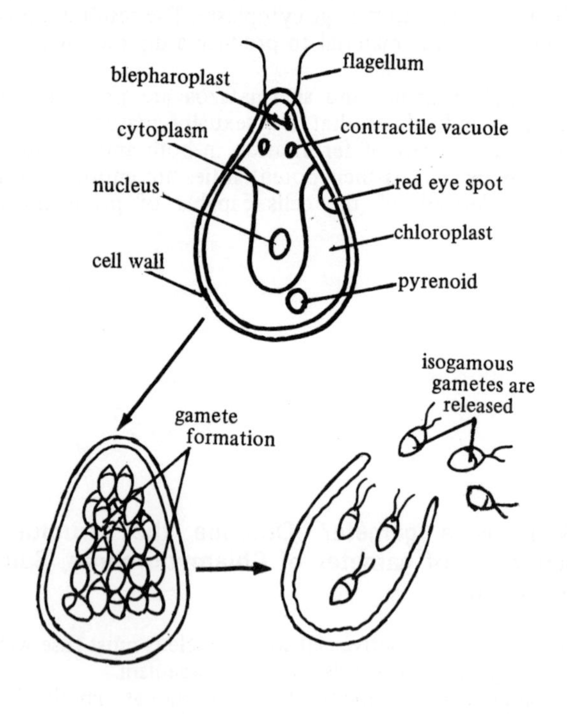

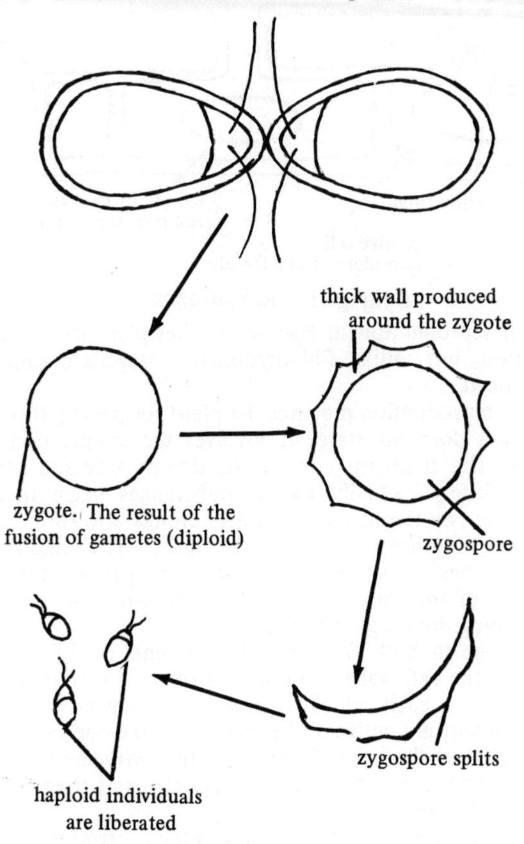

Since all the fusing gametes are identical in size the sexual reproduction is described as isogamous.

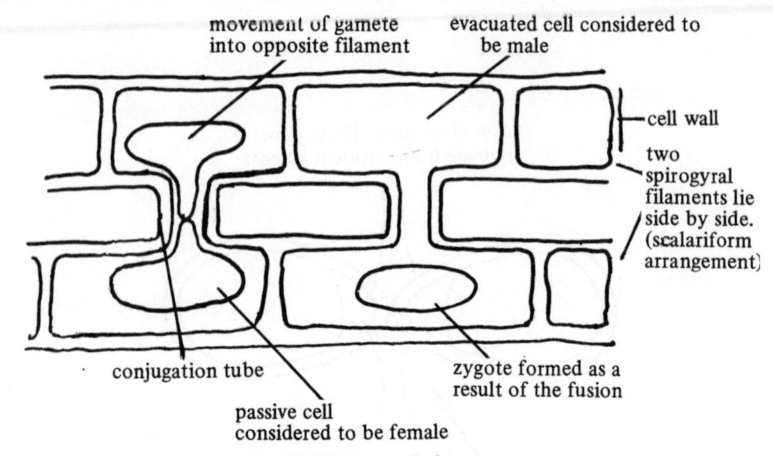

Conjugation in Spirogyra

Sexual reproduction in Spirogyra takes place by a process of conjugation, but unlike Chlamydomanas flagellate gametes are not produced.

Sexual reproduction prepares the plant for passing through the winter in a dormant state. It involves the conjugation of two filaments. The filaments come to lie side by side and adhere by their mucilage. Gradually two protuberances begin to develop between the two filaments and as they elongate to form tubes the conjugating filaments are pushed apart. The filaments are eventually held together by the conjugation tubes. The appearance of the filament is scalariform with the conjugation tubes forming the rungs of the ladder.

Meanwhile in both filaments the contents of the cells shrink away from the cell walls and contract to form oval bodies called gametes, one in each cell. The gametes are isogamous, but those in one filament are active and move across the conjugation tubes into the cells of the other filament to fuse with the passive ones. The active ones are called the male gametes and the passive ones the female gametes.

The fused mass is called the zygote which surrounds itself with a thick cell wall to form a dormant zygospore. When the cell walls of the filament decay the zygospore is released and sinks down into the mud. When favourable conditions return the spore germinates and the thick wall cracks producing a new filament. The new filament is haploid, because the zygotic nucleus divides into four meiotically. Three of these abort and the resulting cell contains one haploid nucleus and hence the resulting filament is haploid.

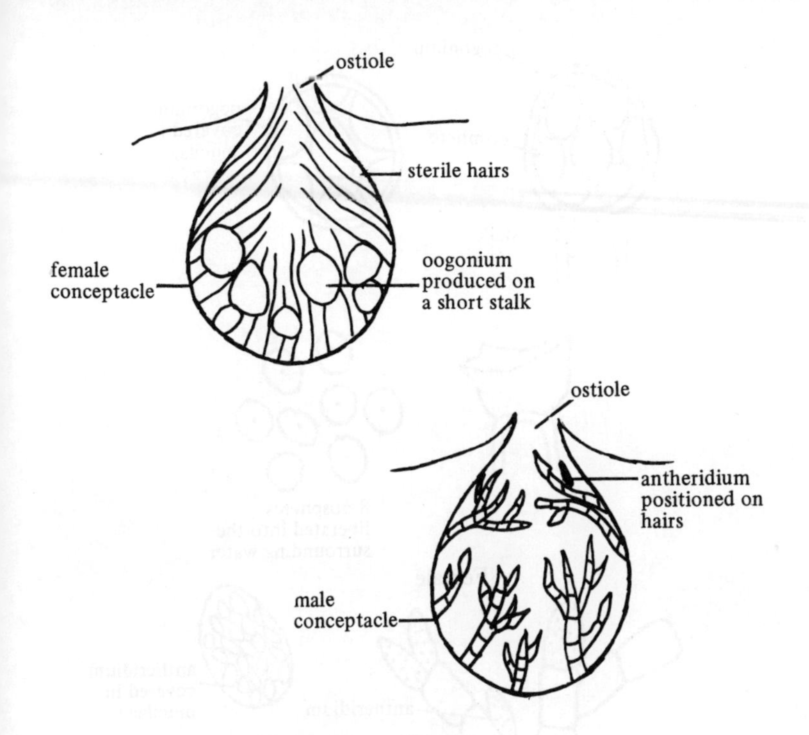

Conceptacles of Fucus

Unlike Chlamydomonas and Spirogyra, Fucus has specialized reproductive structures called conceptacles in which the sex organs are found.

Sexual reproduction in Fucus vesiculosus is dioecious .i.e. male and female plants are formed. The male conceptacles are distinguished by their orange colour while the female are green.

A conceptacle is a hollow flask shaped structure in the thallus, opening to the exterior by a small hole the ostiole. The male conceptacle is richly lined with hairs on which are borne the antheridia. These produce the antherozoids. The female conceptacle also has hairs, but do not bear reproductive organs. The female organs, the oogonia are borne on the walls of the chamber on short stalks among the bases of the hairs. Each oogonium contains eight oospheres, while sixty-four antherozoids are produced in the antheridia. The production of sex organs

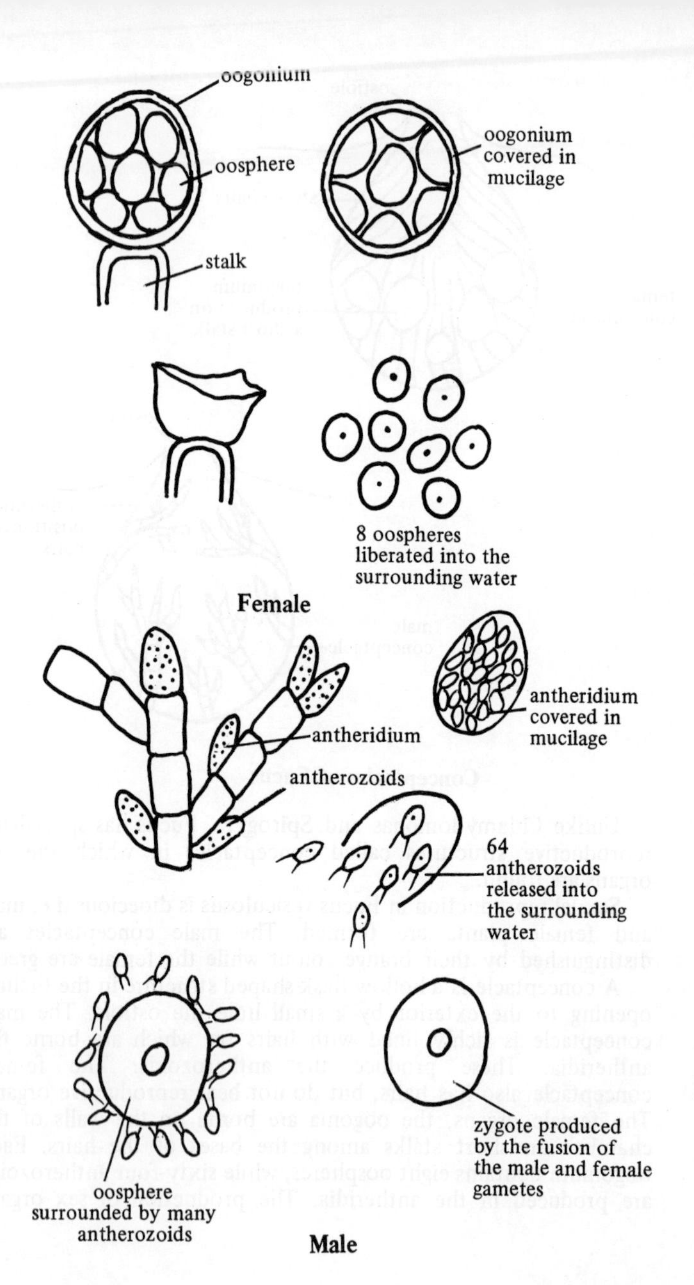
oogonium
oosphere
stalk
oogonium
covered in
mucilage
8 oospheres
liberated into the
surrounding water
Female
antheridium
antherozoids
antheridium
covered in
mucilage
64
antherozoids
released into
the surrounding
water
oosphere
surrounded by many
antherozoids
zygote produced
by the fusion of
the male and female
gametes
Male

produce large quantities of mucilage in the conceptacles.

The mature oosphere is a green spherical body, non-motile, surrounded by a delicate membrane and rich in oil droplets.

When the plants are uncovered at low tide the thalli dry and tend to contract thus squeezing out the contents of the conceptacles. The male and female organs are covered in mucilage and when the tide returns the mucilage is washed off carrying the gametes into the water. The antherozoids and oospheres are liberated into the surrounding water and immediately the antherozoids swim around the oospheres. This is probably the result of a chemical attraction. Fertilization takes place between one antherozoid and one oosphere to form a zygote. The remaining antherozoids die. The zygotic nucleus is diploid, the result of fusion between two haploid cells. The fertilized oosphere gives rise to the new plant, a condition that is not found in Chlamydomonas and Spirogyra i.e. the vegetative structures are haploid.

Sexual reproduction in Fucus is oogamous.

The structure and activities of the gametes in Chlamydomonas, Spirogyra and Fucus all differ, but their function is unchanged in all three and that is to fuse with a gamete of a similar species to produce a new generation of plants.

21. Describe the structure and life history of Mucor.

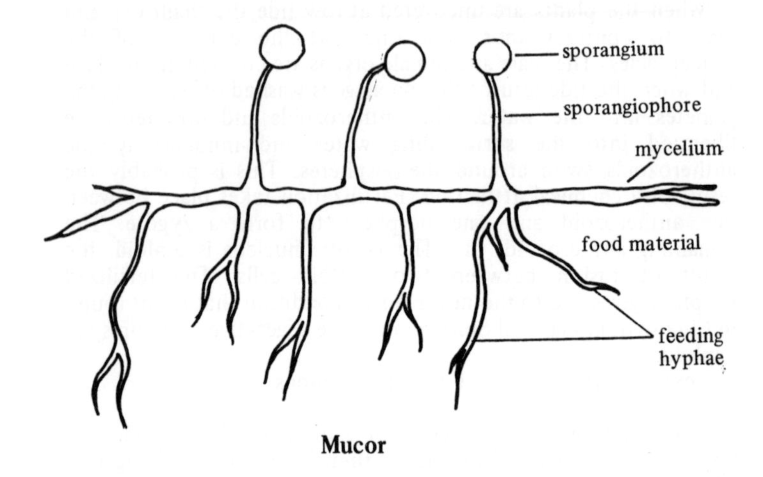

Mucor

Mucor is a saprophytic fungus living on a variety of food stuffs e.g. bread or jam. The mycelium grows in warm damp dark conditions and can be obtained in the laboratory by moistening bread and leaving it in a warm place for a few days.

The mycelium is composed of a mass of white threads with black dots. The white threads branch profusely and are called hyphae. Each hypha is tube-like with a vacuole running through the centre, surrounded by a layer of cytoplasm in which are numerous small nuclei and granules of food material in the form of oil droplets. There are no cross walls in the hyphae and is therefore a coenocytic structure.

There are two types of hyphae. Some lie flat upon the food material and penetrate the food on which the fungus is living. These are the feeding hyphae. Other hyphae grow up into the air. These are called the sporangiophores and bear spherical structures at their apices called sporangia. Within the sporangia are produced numerous small structures called spores.

There are two methods of reproduction, asexual and sexual reproduction.

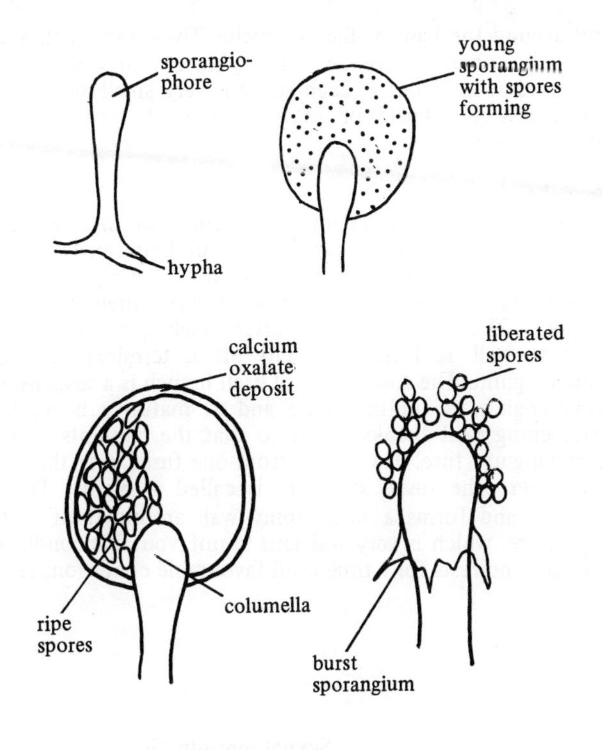

Asexual reproduction

Asexual reproduction

A sporangiophore grows into the air and the tip becomes filled with cytoplasm. A curved cross wall cuts the tip off from the rest of the hypha which becomes the sporangium. This cross wall pushes up into the sporangium and is called the columella. The contents of the sporangium are divided into a mass of uniculeate portions each of which develops a clear resistant wall. Each structure is a spore. The wall of the sporangium becomes much thicker and is impreganted with calcium oxalate which gives it a black colour. The columella becomes very large filling a large space inside the sporangium

Water is absorbed by the sporangium and this sets up a pressure together with the increasing size of the columella within the sporangium and as a result the outer wall disrupts and forms a frill around the base of the columella. The spores are liberated as a closely adhering mass and on drying out they separate to be dispersed as individuals. Each spore is very small and is easily air borne. If a spore settles on a suitable substrate it can germinate to form a new mycelium.

Sexual reproduction

This occurs when conditions become unsuitable for active growth. A sexual union can only occur between hyphae which differ from one another physiologically. The tips of two short hyphae grow towards one another until their tips come in contact. These progametangia after touching gradually swell and a cross wall is formed to cut off a terminal cell called a gametangium. The lower part of each branch is a suspensor. Each gametangium is multinucleate and as maturity is reached the intervening wall breaks down so that the contents of the two gametangium fuse. The nuclei from one fuses with the nuclei of the other. The fused structure is called a zygote. The zygote enlarges and forms a thick spiny wall around itself to form a zygospore, which is very resistant to unfavourable conditions e.g. drought and rests for a time until favourable conditions return.

Sexual reproduction

progametangia

gametangia

hyphae
from two different
physiological individuals

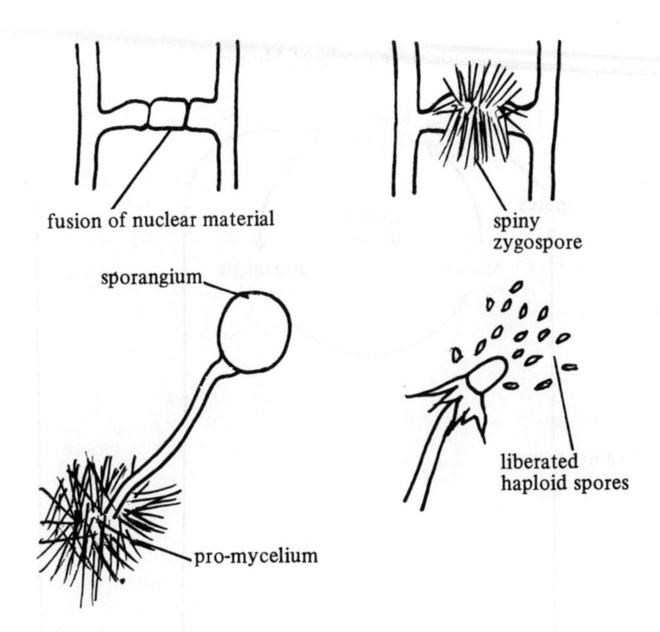

Once it is deposited on a suitable substrate germination is rapid. The zygospore wall bursts and a new hypha develops, this is a pro-mycelial hypha. It gives rise to a single sporangium which develops spores exactly as in asexual reproduction. These spores contain nuclei from the zygote by meiotic division and are therefore haploid. The spores produced by the sporangium are either plus or minus showing that the differentiation of the strains must have occurred in some nuclear division of the zygospore prior to its germination. It is only when the spores from the pro-mycelial sporangium germinate is the life cycle of Mucor completed.

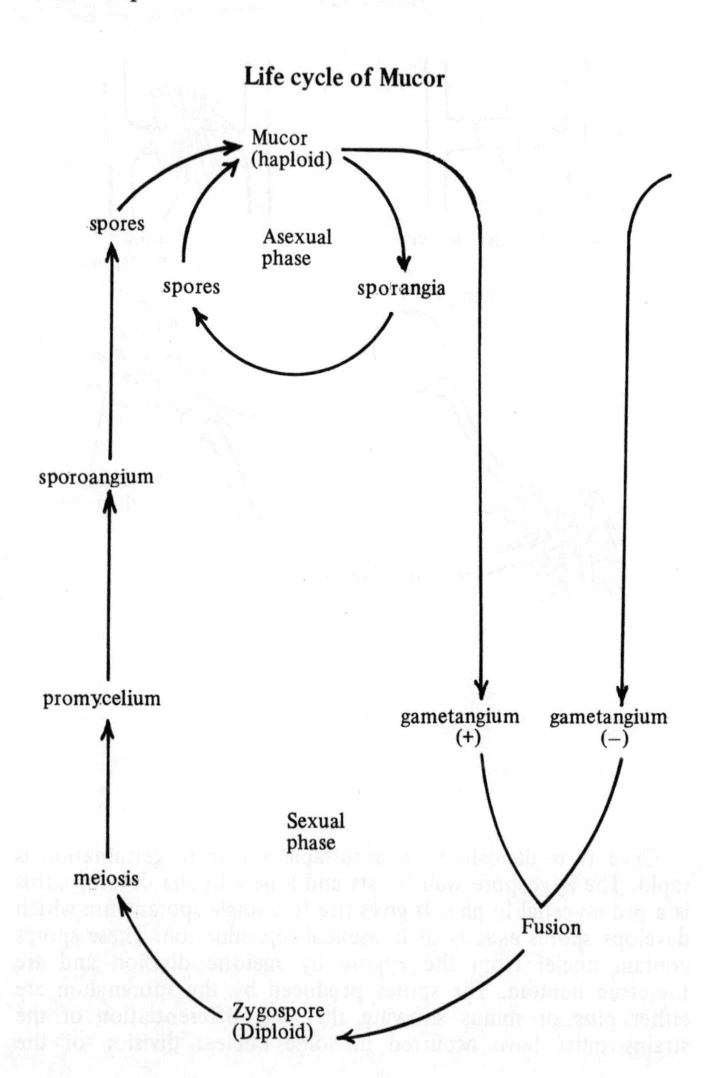

22. Distinguish between an oogonium and an archegonium referring to named examples. Which group of plants possess archegonia? Indicate any biological advantages from the possession of archegonia.

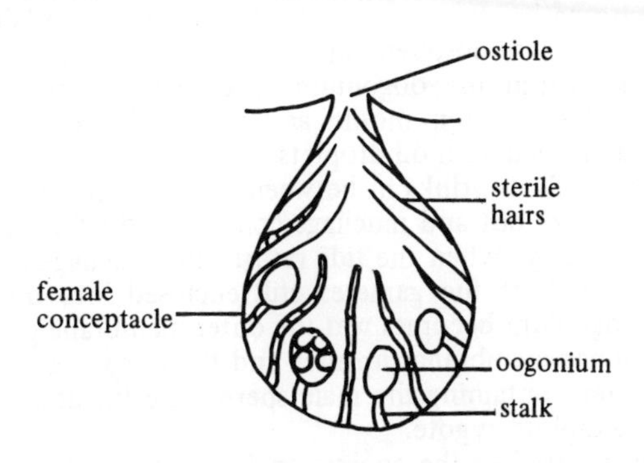

Female conceptacle

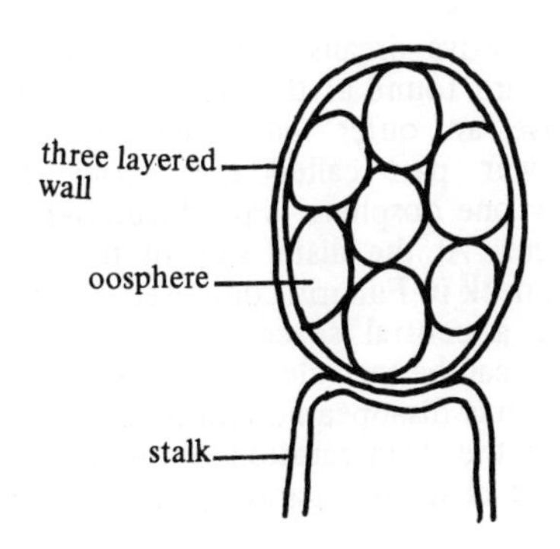

An oogonium enlarged

Reproductive structures called conceptables occur in Fucus vesiculosus. The female conceptacles resemble hollow flasks in the tissues of the thallus, opening to the exterior by an ostiole. The female conceptacles are lined with hairs and among the hair bases may be found the female sex organs the oogonia. They are borne directly on the walls of the chamber on short stalks. Each oogonium has a three layered wall and contains eight oospheres.

While producing new oogonia the conceptacles are filled with mucilage. As an oogonium matures, the nucleus undergoes reduction division followed by mitosis to produce eight haploid nuclei. Each one becomes surrounded by cytoplasm and develops as a oosphere. These oospheres are squeezed against each other as they enlarge within the oogonium. The mature oospheres are green round bodies, non motile and surrounded by a delicate membrane and filled with oil droplets.

When the thallus shrinks in between tides the contents of the conceptacles ooze out and mucilage containing the female organs can be seen clearly. When the tide returns the mucilage is washed off carrying with it the gametes still enclosed into the water. When the oogonium becomes wet the outer membrane gelatinises whilst the inner membrane ruptures and the oospheres are freed into the water containing the male sperms. Fertilization follows to produce a diploid zygote.

Fucus is adapted to the aquatic environment, making full use of the changing wet and dry conditions for the release of gametes into the water.

The female reproductive organs of the Bryophytes are called archegonia and they are found at the apices of fertile branches. The archegonia have an outer wall enclosing a developing oosphere in the lower part called the venter. It must be emphasised that only one oosphere is produced here while eight are produced in Fucus. At the distal side of the venter can be found a long twisted neck in Funaria composed of a series of tiers of cells surrounding a central space or neck canal cells. At maturity the oosphere can be seen clearly through the venter wall. The ventral canal cell has disappeared and the neck is open along its full length due to the disintegration of the neck canal cells. Only a mucilage filled channel separates the oosphere from the exterior.

The archegonium is composed of cells functioning only to protect the female gamete and to form a direct channel through which fertilization can take place. The archegonia are formed from cells just behind the thallus tip. The ridge on which it grows is built up and an involucre grows out from the top of this ridge to cover the depression made.

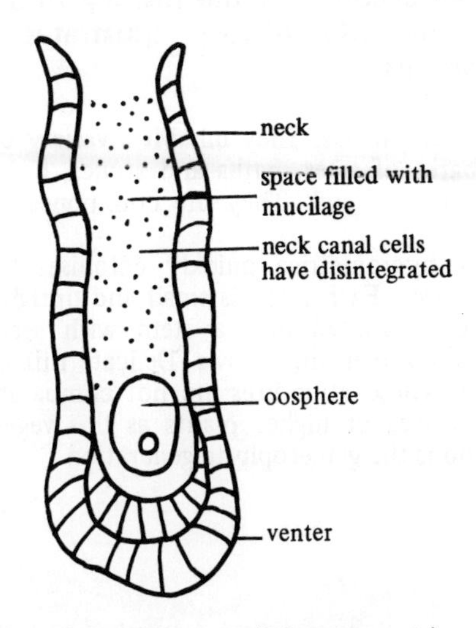

Archegonium of Funaria (Diagrammatic)

The oospheres are non-motile and so a motile antherozoid must seek out the oosphere. This can only be done in wet conditions and an antherozoid must penetrate the neck of the archegonium. The neck of the archegonium secretes a protein substance which attracts the antherozoid. The fusion of the gametes in the venter of the archegonium forms a diploid zygote which will produce the sporophyte generation of the plant.

The Bryophytes and the Pteridophytes e.g. Funaria and Dryopteris produce archegonia.

The biological advantages from possessing archegonia are that the oosphere lies within the structure and is protected by the surrounding tissues from desiccation.

The single egg produced is large and richly supplied with food and therefore there is less drain on the parent plant. The oosphere is stationary which increases the chances of a motile antherozoid reaching it for fertilisation to occur.

23. Give an account of the history of a moss and show how this life history illustrates alternation of generations.

Mosses are gregarious. They inhabit a variety of places e.g. burnt ground, barks of trees, paths and stones, but usually found on damp patches of soil. They are commonest where moisture is plentiful.

Funaria hygrometrica quickly colonises the recent sites of woodland fires. Each plant is small and upright. The plant body can be differentiated into a stem with sessile leaves spirally arranged along it in three rows. Delicate rhizoids grow from the stem base. These structures are not comparable with the stem root and leaves of higher plants as this vegetative structure is haploid and is the gametophyte generation.

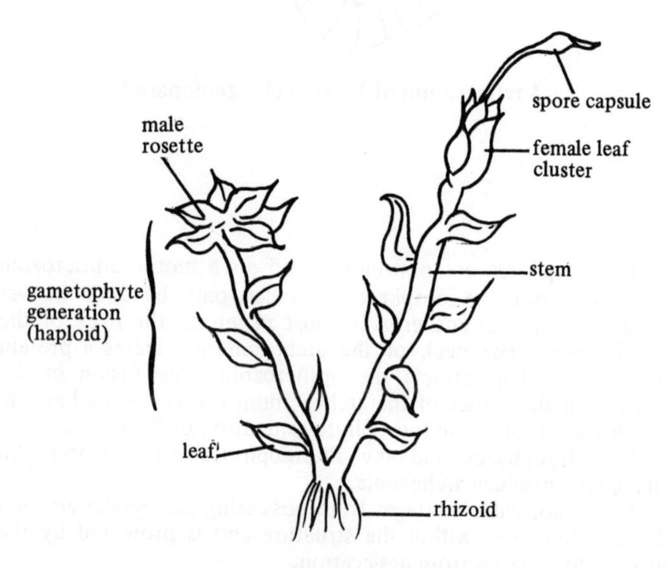

Funaria hygrometrica habit

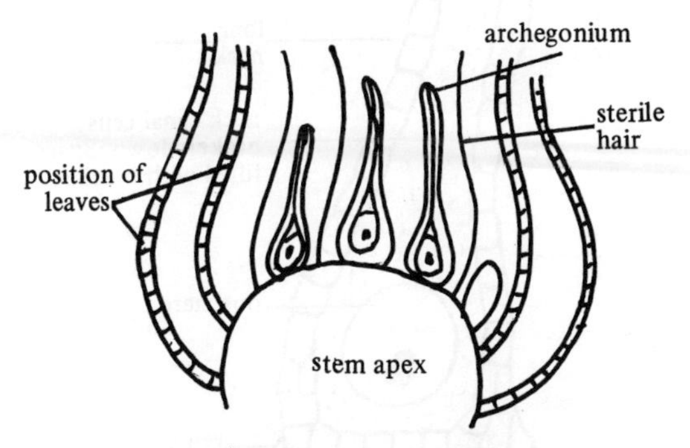

Female apex of thallus

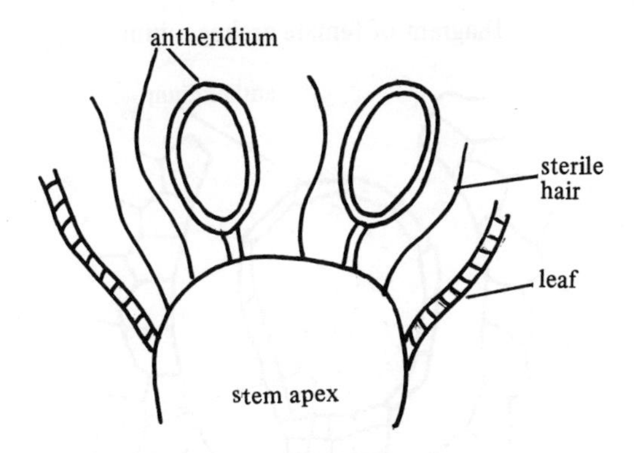

Male apex of thallus

The reproductive organs of Funaria hygrometrica occur on separate sex organs. The male occur at the apex of the main stem and the female organs occur at the apex of a lateral branch. The male axis bears its leaves in an open rosette formation with a brown centre, while the leaves of the female axis forms a cluster around the organs, thus providing them with protection.

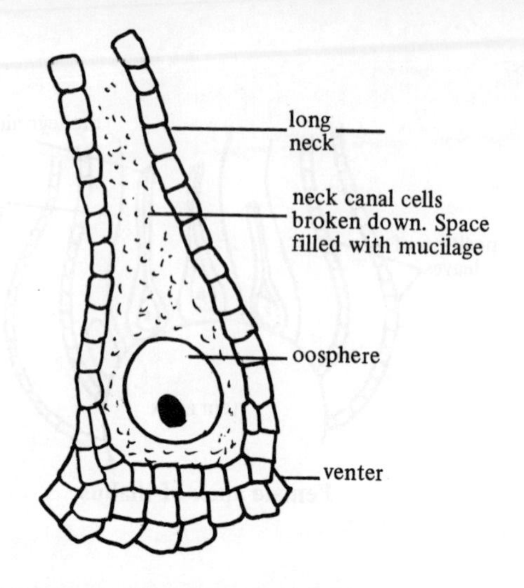

Diagram of female archegonium

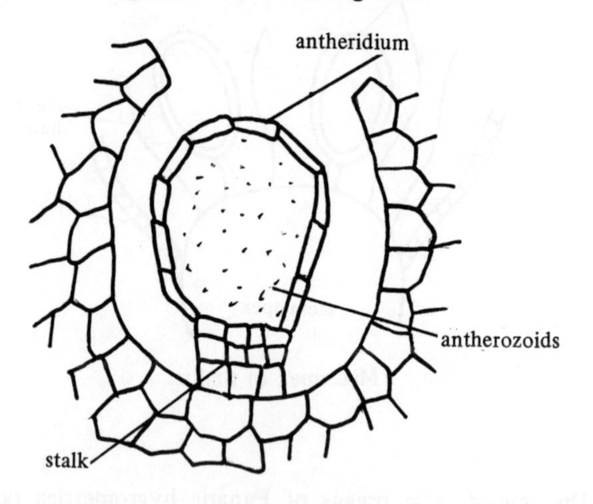

Diagram of male antheridium

The female sex organ is the archegonia. It is composed of a series of sterile cells functioning to protect the female gamete from desiccation or injury and forms a channel through which fertilization can take place. The archegonium arises from cells lying in groups just behind the apex of the thallus. The ridge on which they are formed is built up by the surrounding tissue behind and an involucre grows out from the ridge to cover the depression made. All the divisions are mitotic. The archegonium develops finally and has an outer wall enclosing a developing oosphere in the lower portion called the venter. A neck develops from the venter. The neck is at first closed and contains neck canal cells. Later these neck canal cells break down and the neck opens. Then only mucilage fills the channel and it is only this that separates the oosphere from the outside.

The antheridia are embedded in small outgrowths at the apex of the main stem. Each antheridium is a round mass of cells. The outer layer functions as a protective covering to the inner more fertile layer. Each grows from a single cell which divides mitotically to form the antherozoid mother cells. Later each antherozoid mother cell will give rise to a single antherozoid with two flagellae. The mature antherozoids lie freely in the cavity of the antheridium as the mother cells break down. They are liberated as motile male gametes when the antheridium absorbs water and bursts.

Sexual reproduction occurs. A motile biflagellate antherozoid seeks out and fuses with the non motile oosphere. Moisture must be present to enable the antherozoid to penetrate the neck of the archegonium. The neck of the archegonium secretes a proteinous substance which attracts the antherozoid. Fusion takes place in the venter to form a diploid zygote which becomes the oospore.

This oospore after resting begins to develop and produces a sporogonium which is composed of a capsule, seta and foot.

The oospore divides transversely into two segments. Each segment cuts off a triangular apical cell which proceeds to develop a central elongating portion. The lowermost cells form the foot which grows back into the gametophyte tissue where it establishes contact with the conducting tissue. Thus the sporogonium is dependent on the gametophyte for food and anchorage.

The middle region becomes the seta which conducts food and water to the upper region and it also has mechanical tissue for support. The outer cells become green.

The upper region enlarges and elongates away from the parent. The venter and neck of the archegonium develops increasing in diameter and length to form a calyptra. On further elongation of the seta the calyptra breaks around the base and remains as a cap over the developing capsule, which when ripe is pendulous with the open end downwards. When the sporogonium is mature and dry the seta is twisted.

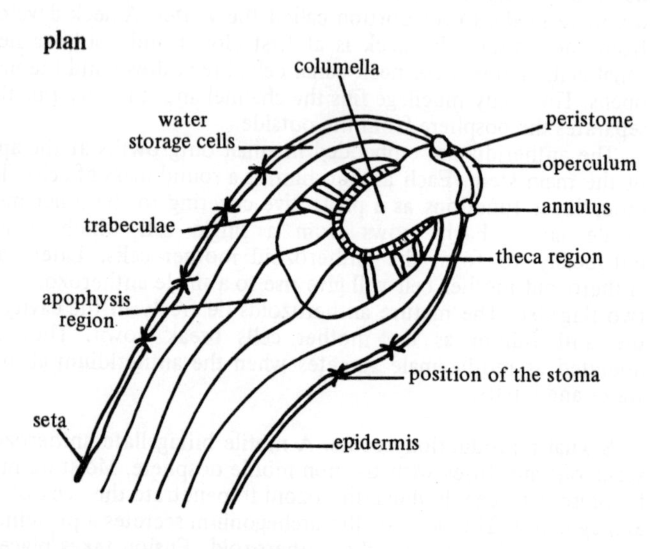

Sporogonium of Funaria

From the diagram of the sporogonium it can be seen that there develops at the apical end sporogenous tissue which is traversed by trabeculae. The upper spore bearing tissue is the theca which has a complicated covering at the apex. There is an inner peristome and an outer operculum. The apophysis and all the outer layers of the theca are formed from the amphithecium, whilst the spores are developed from the outer layers of the endothecium, the columella is the inner portion of this. The reproductive capacity of the sporogonium is purely asexual. When mature the whole structure dries. The columella cells and the inner amphithecial cells surrounding the ripening spores break down to form a hollow spore sac closed at the upper end by the peristome and operculum. Within the space, the spores lie as a powdery mass. Each spore is protected by a smooth exosporium which contains oil and a little chlorophyll. As the capsule dries the calyptra falls away and the operculum is forced off by a movement of the annulus. This peels off and is turned inside out due to the swelling of the innermost layer of cells which have mucilaginous walls. The peristome teeth show hygroscopic movements bending away from the centre when dry and closing again when damp.

The seta bends back and the capsule hangs downwards. In dry conditions the teeth are open and the ripe spores fall out to be carried away by the wind.

The sporogonium is partially self supporting depending on the parent plant for its water and mineral salts. All the spores are haploid division being meiotic.

When a spore reaches a suitable substratum it begins to germinate. The exosporium ruptures and two or three germ tubes emerge. One forms a cross wall and gives rise to a branched filamentous structure the primary protonema. The protonema spreads and eventually forms buds. When an apical cell is differentiated a small moss plant, the gametophyte develops. This then completes the life cycle of the moss which shows an alternation of generations.

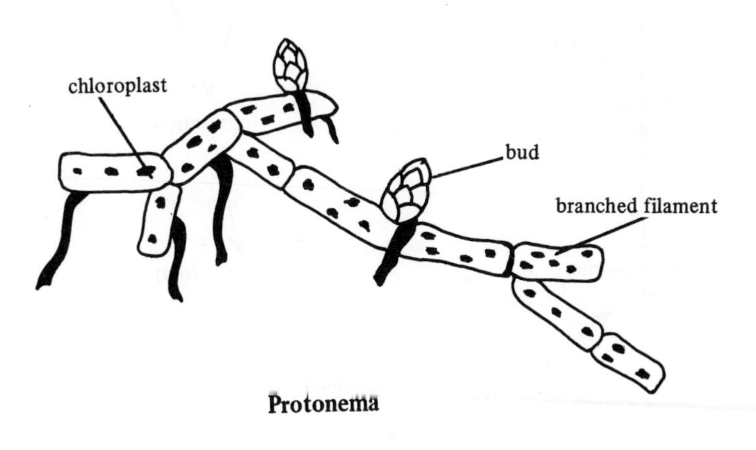

Protonema

Life cycle of Funaria to show alternation of generations

A Summary

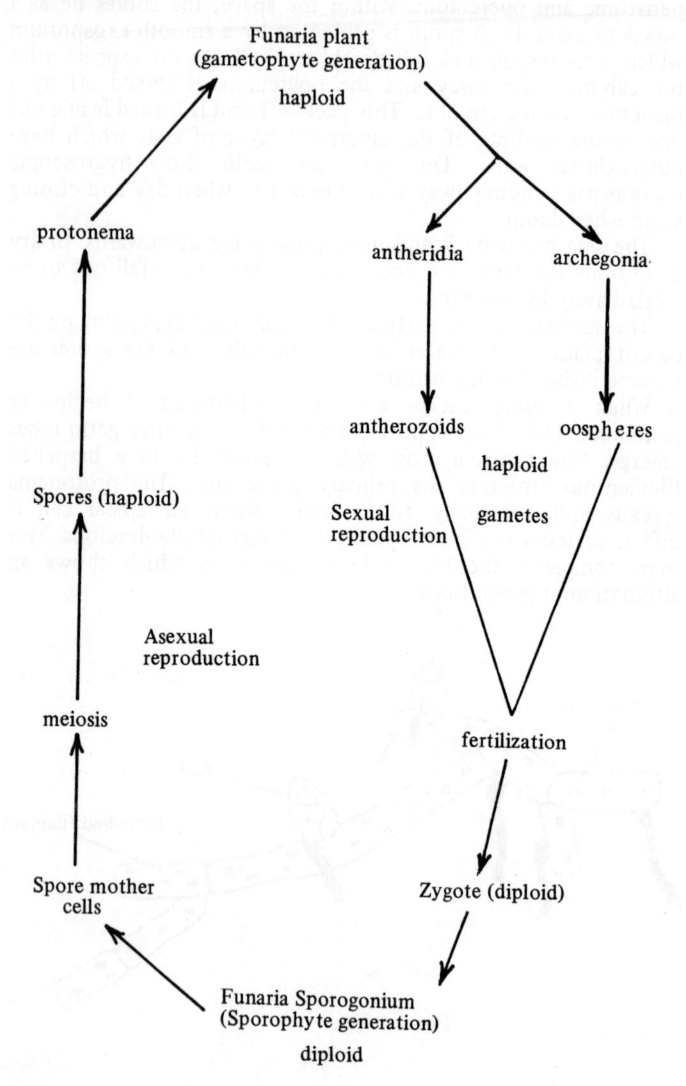

24. Describe what happens when fern spores are sown under suitable conditions. Give an account of the structure and functions of the resulting plants.

When fern spores are sown in the presence of moisture and a suitable temperature they start to germinate and produce a new generation of plants.

Germination starts when the spore wall breaks and a short green filament develops which is attached to the soil by colourless rhizoids. These rhizoids absorb water and mineral salts. The filament of cells continues to divide and widen to form a single sheet of cells called the prothallus, which is the gametophyte generation of the life cycle.

The peripheral cells retain their single layered arrangement but the cells in the central region divide further forming a massive central cushion several cells thick.

The whole body is capable of an independent existence, obtaining nourishment from the soil and photosynthesising. There is a large proportion of surface area to bulk and no resistance offered to the evaporation of water from it in dry air. It is less adapted for life on land than the sporophyte generation and can only survive in damp conditions.

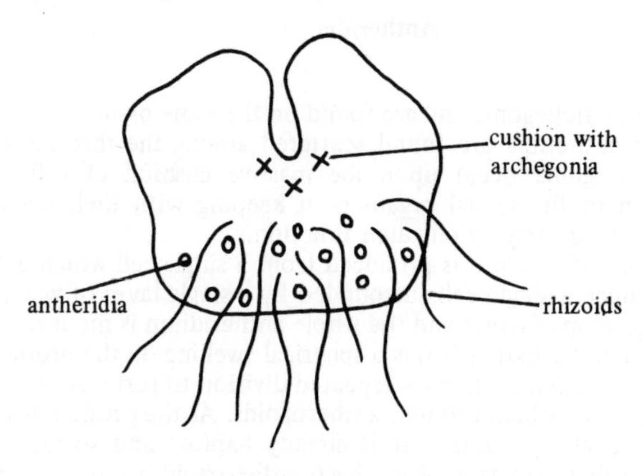

Prothallus of the fern

On reaching maturity the prothallus has developed sexual organs on its undersurface. They are male or antheridia and

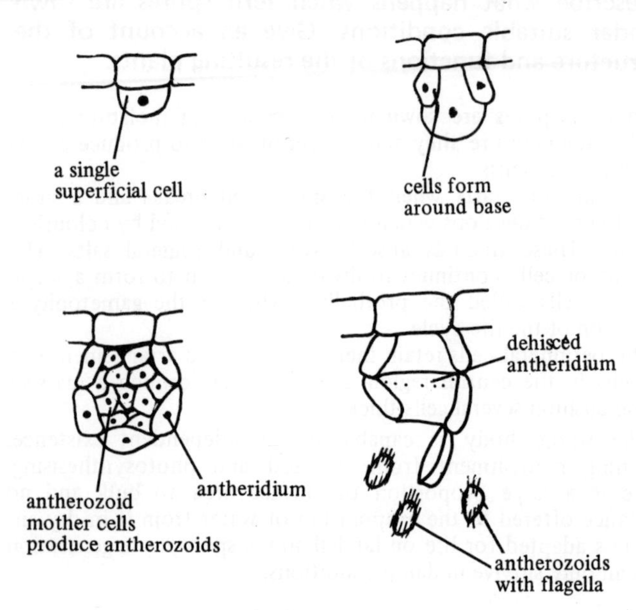

Antheridia

female or archegonia and are found on the same plant.

The antheridia are found scattered among the rhizoids while the archegonia occur upon the massive cushion of cells. The position of the sexual organs is in keeping with their need for moisture for carrying out their function.

Each antheridium is produced from a single cell which divides to produce a single cell surrounded by a single layer of wall cells. This develops further and the whole antheridium is mounted on a short stalk so that it forms a spherical swelling on the prothallus. The central cell undergoes repeated division to form antherozoid mother cells which produce antherozoids. As the prothallus is the gametophyte generation it is already haploid and so reduction division does not take place. Each antherozoid has flagella and is motile. The antherozoids are liberated when the mother cell wall bursts. The cells gelatinise and swell forcing the lid cell of the antheridium. The antheridia are ripe before the archegonia and so self fertilisation does not take place.

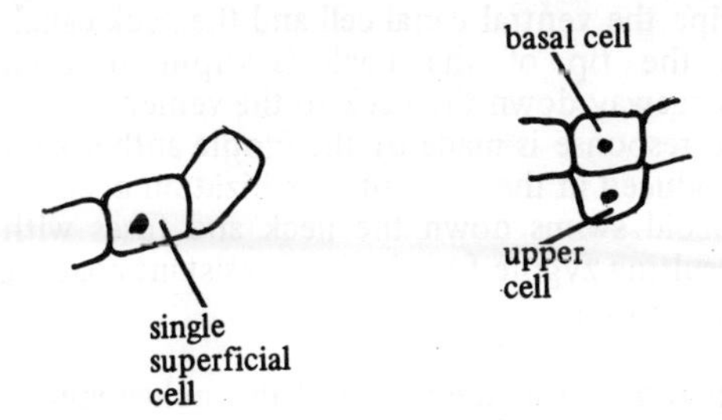

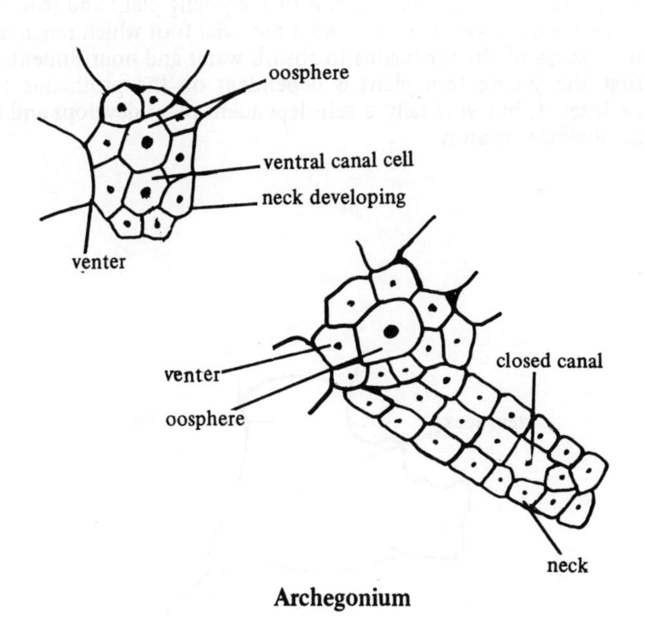

Archegonium

The archegonia also originate from a single cell which divide and gradually produces a lower venter embedded in the tissues of the prothallus. It is the venter that contains the oosphere and above the ventral canal cell. The neck is composed of four rows of cells joined laterally around the central canal. When the archegonium is ripe the ventral canal cell and the neck canal cells disintegrate and the tip of the neck is ruptured leaving a mucilaginous passageway down the neck to the venter.

A chemotactic response is made by the motile antherozoids to the malic acid produced in the neck and fertilization occurs when a motile antherozoid swims down the neck and fuses with the oosphere. The resulting zygote forms a hard resistant coat around itself and forms an oospore.

The oospore is retained in the venter of the archegonium. The zygote divides into a group of eight cells. Four grow towards the apex of the prothallus and four away from it. The first group form the main axis and first leaf of the young plant and from the second group, the first root and a suctorial foot which remains in the tissues of the prothallus to absorb water and nourishment. At first the young fern plant is dependent on the prothallus that encloses it, but gradually a self dependent plant develops and the prothallus rots away.

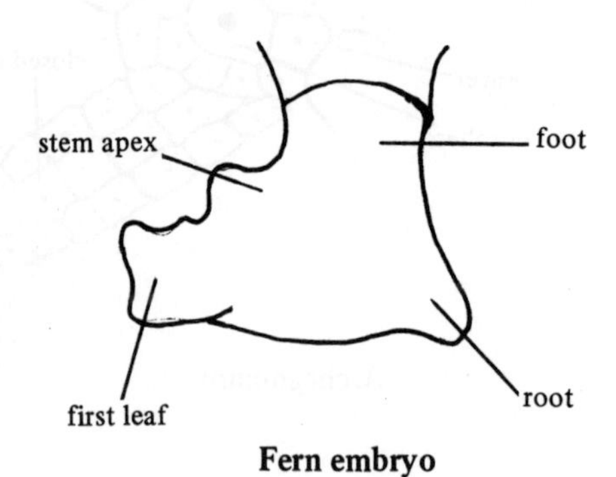

Fern embryo

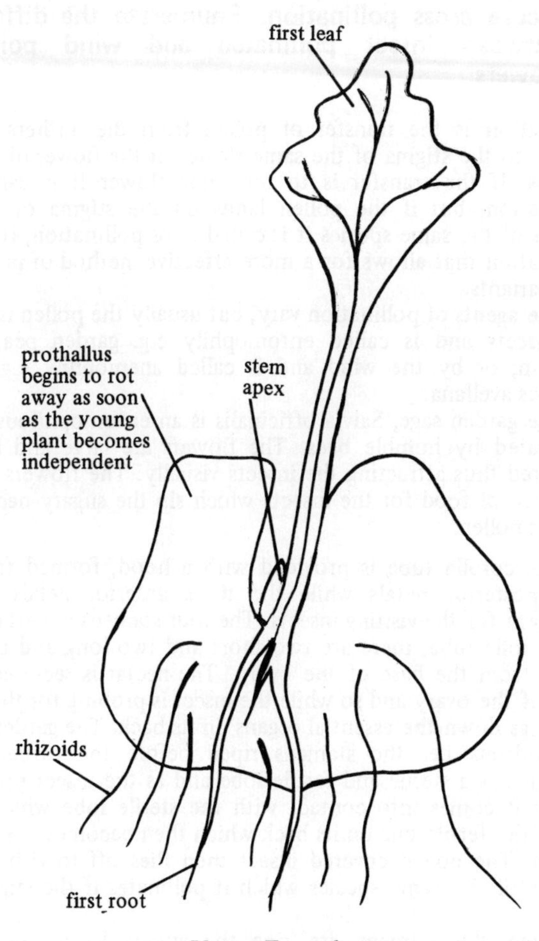

Young Fern plant

25. What do you understand by the term pollination? Describe both an insect pollinated flower and a wind pollinated flower which show special structures to secure cross pollination. Enumerate the differences between insect pollinated and wind pollinated flowers.

Pollination is the transfer of pollen from the anthers of one flower to the stigma of the same flower or the flower of another species. If the transfer is to the same flower it is called self pollination, but if the pollen lands on the stigma of another flower of the same species it is called cross pollination. It is cross pollination that allows for a more effective method of producing new variants.

The agents of pollination vary, but usually the pollen is carried by insects and is called entomophily e.g. garden pea, Pisum sativum; or by the wind and is called anemophily e.g. hazel, Corylus avellana.

The garden sage, Salvia officinalis is an entomophilous flower pollinated by humble bees. The flowers are large and brightly coloured thus attracting the insects visually. The flowers provide a source of food for the insects which sip the sugary nectar and collect pollen.

The corolla tube is provided with a hood, formed from the two posterior petals while the three anterior petals form a platform for the visiting insects. The four stamens are attached to the corolla tube, there are two short and two long and the style arises from the base of the ovary. The nectar is secreted at the base of the ovary and so while the insect is probing for the nectar it brings down the essential organs on its back. The garden sage is protandrous i.e. the stamens ripen before the carpels. Each stamen has a sterile and fertile lobe and as the insect probes for nectar it comes into contact with the sterile lobe which brings down the fertile one on its back which then becomes dusted with pollen. The pollen covered insect then flies off to visit another flower of the same species which it pollinates if the stigmas are ripe.

When the stamens are ripe the stigma lobes are pressed together, but during the female stage the stigma projects thus coming into contact with the pollen on the back of a visiting insect thus ensuring cross pollination. The stamens during this stage bend out of the way or fall off.

Garden sage (protandrous)

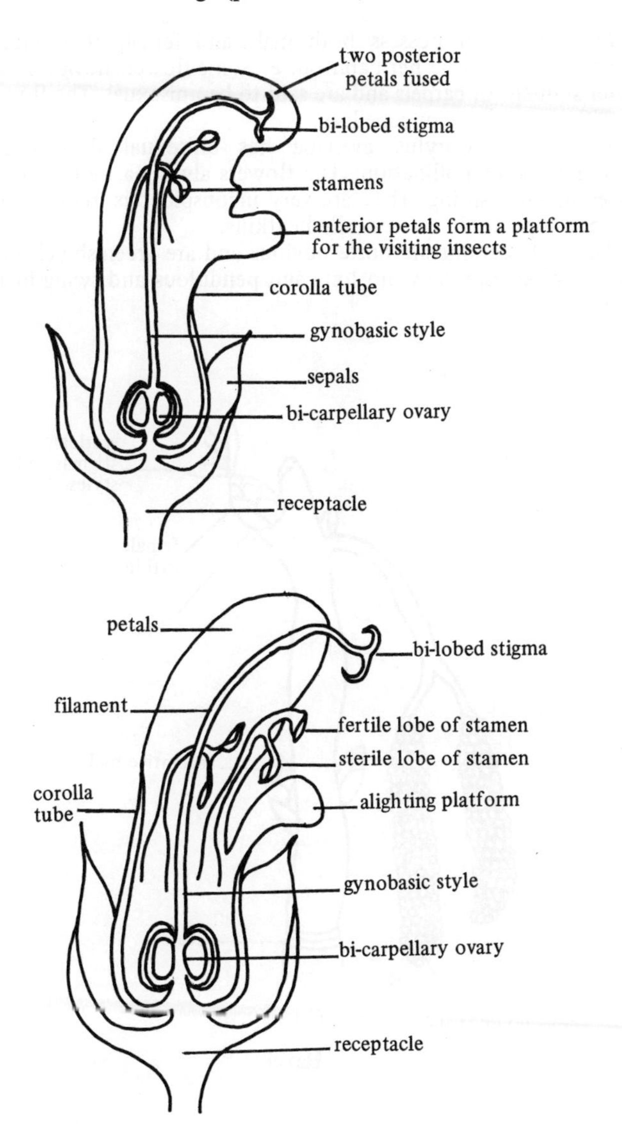

Thus the protandrous condition ensures that cross pollination takes place.

Flowers which possess both male and female reproductive organs are said to be hermaphrodite. Some flowers however have either stamens or carpels and are said to be unisexual. The flowers are either staminate or capellary.

The hazel, Corylus avellana has unisexual flowers and undergoes wind pollination. The flowers always arise before the leaves in early spring. They are very inconspicuous, reduced and grouped together in masses called catkins.

The male catkins are quite obvious and are greenish yellow in colour. When ripe they are long and pendulous and swing in the wind.

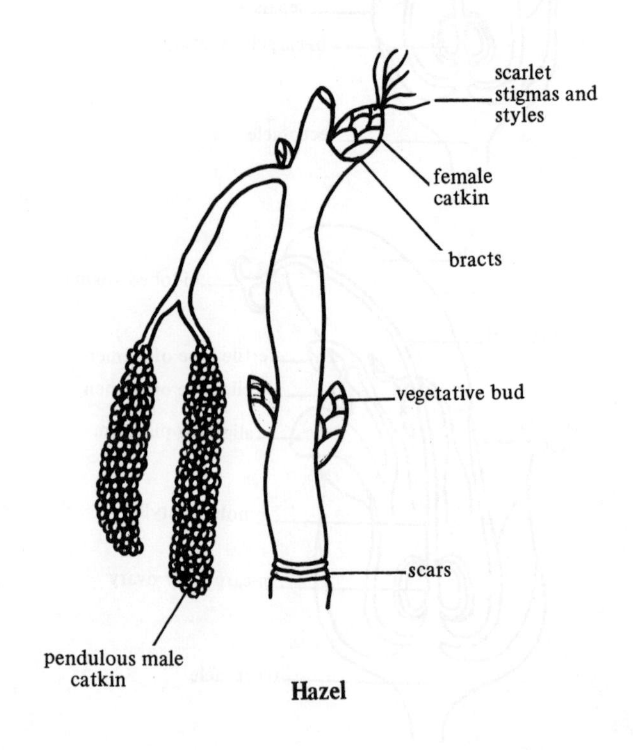

Hazel

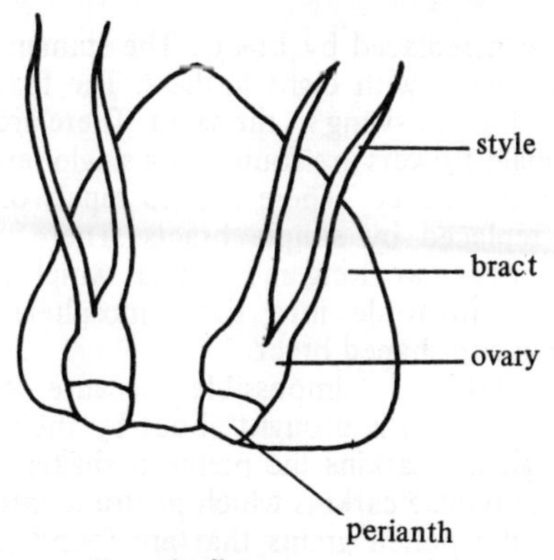

Female flower

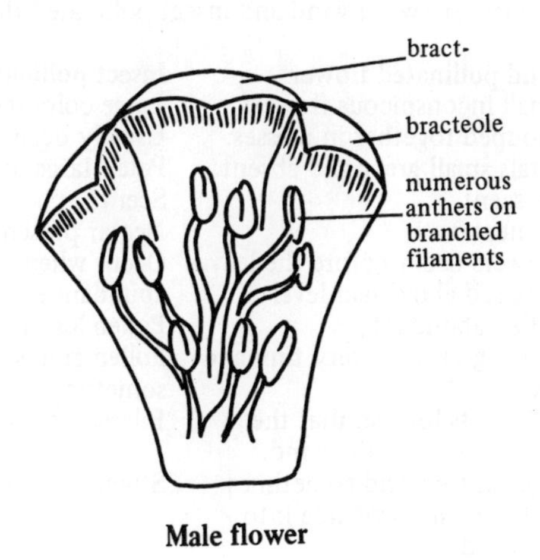

Male flower

The female catkin are much more difficult to detect as they closely resemble an ordinary bud. They are distinguished from these by having red stigmas which protrude from their tips.

Many male flowers are found on a single male catkin. Each flower is very much reduced. There are no sepals or petals for these have been replaced by bracts. The stamens are made up of branched filaments with eight anthers. The filament is long and allows the anthers to swing in the wind. There are no carpels.

Many female flowers are found on a single female catkin. Each flower is again reduced. There are no sepals or petals for these have been replaced by simple bracts. There are two separate carpels each with two long red stigmas which grow out beyond the bract and protrude into the atmosphere. Each carpel is contained in a cup-shaped bract.

Self pollination is impossible because the flowers are unisexual. Pollination is brought about by the wind. As the wind passes through the catkins the pollen is shaken into the air. The stigmas of the female catkins which protrude into the atmosphere can pick up the pollen grains that are found in the air. Pollen distribution is not hindered by the leaves because the flowers are formed before the leaves. Thus cross pollination occurs.

Differences between wind and insect pollinated flowers

	Wind pollinated flowers	**Insect pollinated flowers**
1.	Small inconspicous flowers.	Large coloured flowers.
2.	Grouped together in masses.	Usually occur on their own.
3.	Petals small green and absent.	Petals large and colourful.
4.	No scent.	Scent.
5.	No nectar.	Nectar present.
6.	Flowers occur before the leaves or raised above leaf level	Occur when there is an abundance of insects.
7.	Pollen abundant,	Pollen less abundant.
8.	Pollen grains are very tiny and dry.	Pollen grains larger and sometimes sticky.
9.	Filaments long so that the anthers hang in the wind.	Filaments usually short.
10.	Stigmas long and sometimes feathery and protrude into the wind.	Stigmas short and sticky.

ADDITIONAL QUESTIONS

Additional Questions

1. Describe the structure of Paramecium explaining the uses of its organelles. What advances structurally does it show over Amoeba?
2. Give an account of the ways in which Protozoa move. Indicate wherever possible the relation between feeding and movement.
3. Distinguish between intracellular and extracellular digestion, illustrating your answer by means of the types of animals in your syllabus.
4. Give an account of the increasing complexity of the organs of locomotion as illustrated by the series Amoeba, Earthworm and Periplaneta.
5. What is meant by excretion? Describe how this takes place in Amoeba, Lumbricus and a named mammal.
6. Describe how Fasciola produces large numbers of offspring.
7. Write short notes on the following a) strobilization, b) symbiosis and, c) regeneration. Illustrate your answer by reference to organisms you have studied.
8. Write an essay on metamorphosis in insects.
9. Describe the structure and function of a) the cochlea, b) the retina of a mammal.
10. Describe the histology of the mammalian testis. What functions are ascribed to this organ?
11. Give a diagram of a uriniferous tubule and discuss its functions. Distinguish between excretion and osmo-regulation.
12. Describe and illustrate the arrangement of the embryonic membranes of a named mammal.
13. Enumerate the differences between arteries and veins. With the aid of fully labelled diagrams trace the path of a red corpuscle from a) the hind leg of a mammal to the lungs, b) from the head to the liver and c) from the right auricle to the kidneys.
14. Define excretion? Make a large fully labelled diagram of the skin of a mammal. What are the functions of each part you label?
15. Write an essay on either the secretions of cells or sense organs.
16. The muscles of the rat's hind leg requires oxygen. Explain

how the oxygen reaches them.

17. Describe the synthesis, storage and digestion of carbohydrates in a) flowering plant and b) mammal.
18. By reference to the structure and physiology of Euglena, explain its importance as a link between plant and animal kingdoms.
19. What do you understand by a Fungus? Illustrate your answer by reference to any Fungus you have studied.
20. Discuss the role of bacteria in nature.
21. How would you classify Chlamydomonas, Mucor and Spirogyra? Give reasons for your answer.
22. Describe and compare the structure of the thallus in Mucor and Spirogyra. Comment on the nutrition and reproduction of these two organisms.
23. Give an account of the structure and life history of Fucus. How is the plant adapted to its habitat?
24. Define the terms gamete, gametangium and zygote? Describe these structures in a named alga and a named fungus.
25. Give an account of the structure and life history of a named moss. To what extent do you think that mosses are adapted to a terrestial habit?

CELTIC REVISION AIDS

An extensive range of study and revision material which may be used by students while preparing for a wide range of examinations. The material is designed such that the student can use on his own and requires no supervision or guidance. It can be used equally well in a classroom or in the student's own home. The material can be used as part of a programmed revision course or as a last minute 'brush-up' on essential facts and examination techniques.

Series available are:

Rapid Revision Notes
A series of books designed for use by students studying GCE 'O' level, CSE or similar examinations. The essential facts of the subject are reviewed in note form and typical examination questions are given at the end of each section. These notes are organised such that the essentials of the subject can be reviewed and revised as simply as possible.

Subjects covered in this series are: English Language Book 1 and Book 2, Commerce, Economics, Sociology, British Economic History, Physical Geography, Mathematics, Chemistry, Physics, Biology and Human Biology.

Model Answers
A series aimed at GCE 'O' Level, CSE, RSA and 16 plus level examinations. Typical examination questions are presented and suggested answers are given. This series helps students remember essential facts and the best methods of presenting them in examination conditions.

Subjects covered in this series: Julius Caesar, Macbeth, Romeo and Juliet, The Merchant of Venice, Essay Writing, Precis Writing, Mathematics, Physics, Chemistry, Biology, Human Biology, Commerce, Economics, British Isles Geography, British Economic History, Accounts and Commercial Mathematics.

Worked Examples
A series with the same basis as the Model Answers series, but aimed at the GCE 'A' Level and similar examinations.

Subjects covered in this series are: Pure Mathematics, Applied Mathematics, Statistics, Chemistry, Physics, Biology, Economics, Sociology, British History, European History, British Economic History, Physical Geography and Accounts.

Multiple Choice 'O' Level
Multiple choice questions are a very important part of the examination requirements for GCE 'O' Level and CSE. This series provides batteries of common questions and is also a very good way of revising essential facts.
Subjects covered in this series are: English, French, Mathematics, Modern Mathematics, Commercial Mathematics, Chemistry, Physics, Biology, Human Biology, Commerce, Economics, British Isles Geography and Accounts.

Multiple Choice 'A' Level
Objective tests are now an important part of most 'A' level examinations. This series presents batteries of common questions and is also an excellent way of revising essential facts.
Subjects covered in this series are: Pure Mathematics, Applied Mathematics, Statistics, Chemistry, Physics and Biology.

Test Yourself
A series of pocket books designed for the revision of essential facts whenever the student has a free moment.
Subjects covered in this series are: English Language, French, German, Commerce, Economics, Mathematics, Modern Mathematics, Commercial Mathematics, Statistics, Chemistry, Physics, Biology, Human Biology, Accounts, British Isles Geography, British Economic History, St Matthew, St Luke, St John, St Mark and Acts of the Apostles.

Celtic Revision Aids can make the difference between passing or failing your examination.